INTEGRATED WATERFLOOD ASSET MANAGEMENT

INTEGRATED WATERFLOOD ASSET MANAGEMENT

by

Ganesh C. Thakur, Ph.D., MBA
Manager—Exploration and Production Operations Services
and Senior Principal Consultant—Reservoir Management
Chevron Petroleum Technology Company
La Habra, California

Abdus Satter, Ph.D.
Senior Research Consultant
Texaco E&P Technology Department
Houston, Texas

Contents

Foreword

The book *Integrated Waterflood Asset Management* fills an important gap for all those asset team members who are responsible for managing water injection projects in an integrated way. This book continues to build on the theme of integrated petroleum resource management, developed by a previous contribution of the authors, *Integrated Petroleum Reservoir Management*. There is, of course, more focus in the present book on the special considerations that integrated teams can provide to optimize the profitability of water injection projects.

Waterflood management has been an important issue for several decades. Although the subject of waterflood management is not new, we continue to learn how to manage waterfloods better each year with our enhanced technological capabilities. When I entered the petroleum industry in the early 1960s, several of my early assignments dealt with waterflood design, working as a reservoir engineer with geologists, facilities engineers, and drilling engineers. Many of the techniques that I used then are still mainstream technology, including reservoir description, followed by reservoir simulation to study alternatives. But what a difference in how the specialists work together! My geologist friends handed their maps off to me; I handed my injector and producer locations and expected rates off to my facilities and drilling engineering comrades – and totally different individuals implemented the plans and followed the progress of the floods. Many of my communications and requests had to go through my boss, the district reservoir engineer, to his peer, the district production geologist, back to my peer, the geologist working on the field.

As this book points out so clearly, waterflood projects are not usually managed this way in the late 1990s. We now serve on integrated teams, with considerable feedback to all team members whenever any member obtains new data – for an infill well, or from a producer with premature water breakthrough, for example. Communications are now streamlined and simplified, action is much more immediate, and profitability is almost always enhanced.

The authors share their personal knowledge and experience in this book, in addition to providing concise surveys of recommendations from the petroleum literature on how to manage waterfloods with these integrated teams. The historical lessons we have learned about waterflood design, reserves estimating, surveillance, field operations, and project economics are all here – but presented the modern way, implemented through team efforts.

This book, like its predecessor, *Integrated Petroleum Reservoir Management*, will serve the industry well.

W. John Lee,
Peterson Chair and Professor,
Texas A&M University

ix

Acknowledgments

The authors wish to acknowledge the support and permission of Chevron and Texaco to publish this book. We also would like to thank our many coworkers and students from whom we have learned about various aspects of waterflood asset management. In particular, we acknowledge the contributions of the following persons for providing certain materials; Wayne Subcasky and John Bagzis for field operations, Chuck Magnani for surveillance and monitoring, Gary Burkett for waterflood design, Jim Baldwin and Rich Jespersen for computation, and John Dacy for assisting with laboratory data on waterflood. Our special thanks go to Joyce Bube for her commitment, patience and hard work in preparing the manuscript. Last but not the least, we owe sincere appreciation and thanks to Pushpa Thakur and our families for their patience, understanding, and encouragement.

Ganesh C. Thakur   *Abdus Satter*

List of Figures and Tables

CHAPTER 6 CAPTION

CHAPTER 12 CAPTION

1

Introduction

Primary methods using natural producing mechanisms, i.e., liquid and rock expansion and solution gas drive, leave behind 80% or more of the original oil in place. Figure 1-1 and Table 1-1 present some characteristics of natural producing mechanisms of oil reservoir. Consequently, a vast amount of hydrocarbon remains unrecovered in the United States and elsewhere in the world.

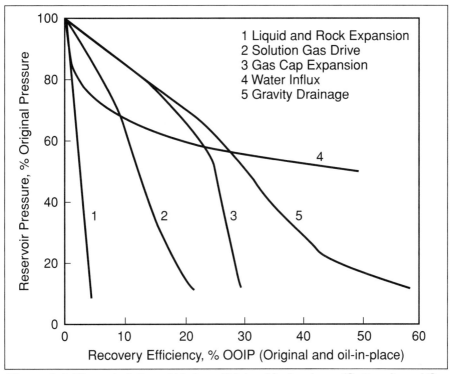

Figure 1-1 Characteristics of Various Driving Mechanisms (Satter, A. and G. C. Thakur. "*Integrated Petroleum Reservoir Management: A Team Approach*," PennWell Books, Tulsa, OK, 1994)

Table 1-1 Characteristics of Various Driving Mechanisms (Satter, A. and Thakur. G. C. "Integrated Petroleum Reservoir Management: A Team Approach," PennWell Books, Tulsa, OK, 1994)

Mechanisms	Reservoir Pressure	GOR	Water Production	Efficiency	Others
1. Liquid and rock expansion	Declines rapidly and continuously P_i (initial pressure) > P_b (bubble point pressure)	Remains low and constant	None (except in high S_w reservoirs)	1-10% Avg. 3%	
2. Solution gas drive	Declines rapidly and continuously	First low, then rises to maximum and then drops	None (except in high S_w reservoirs)	5-35% Avg. 20%	Requires pumping at an early stage
3. Gas cap drive	Falls slowly and continuously	Rises continuously in up-dip wells	Absent or negligible	20-40% Avg. 25% or more	Gas breakthrough at a down dip well indicates a gas cap drive
4. Water drive	Remains high. Pressure is sensitive to the rate of oil, gas, and water production	Remains low if pressure remains high	Down-dip wells produce water early and water production increases to appreciable amount	35-80% Avg. 50%	N calculated by material balance increases when water influx is neglected
5. Gravity drainage	Declines rapidly and continuously	Remains low in down-dip wells and high in up-dip wells	Absent or negligible	40-80% Avg. 60%	When k > 200 mD, for mation dip > 10° and μ_o low (< 5 cP)

Billions of barrels of additional oil have been recovered through waterflooding, which is the most important method for improving recovery from oil reservoirs. With the uncertainty of the applicability of EOR techniques due to oil price instability, integrated waterflood asset management has become more significant than ever. It is not restricted to an initial engineering and geologic evaluation, economic justification, and project approval by management. Rather, these ongoing activities span the time before the start of the waterflood to the time when the secondary recovery either is uneconomic or is changed to enhanced recovery.

Integrated waterflood asset management consists of reservoir characterization, fluids and their behavior in the reservoir, creation and operation of wells, and surface processing of the fluids.[1] These are inter-related parts of a unified system. Figure 1-2 illustrates this unified system. In the past, attention was focused mainly on reservoir performance. However, with the application of the reservoir management approach, it has become industry practice to include wells, facilities, water system, and field operations in waterflood asset management.

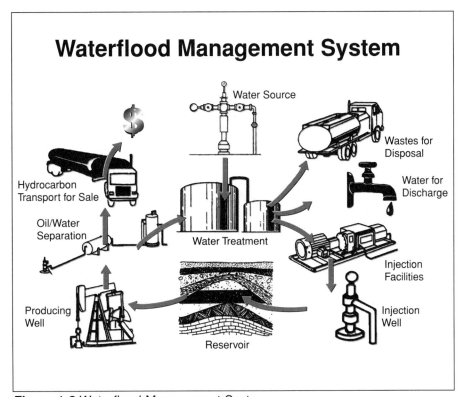

Figure 1-2 Waterflood Management System

This chapter presents:

- Reasons for waterflooding
- History of waterflooding
- Importance of integrated waterflood asset management
- The state of the art
- Scope and objective of book
- Organization of book

Reasons for Waterflooding

Waterflood is a widely used commercial recovery process for the following reasons:

- The water is generally available
- Water is an efficient agent for displacing light/medium gravity oil
- Flooding involves low capital investment and operating costs and has favorable economics
- Water is relatively easy to inject
- Water spreads easily through an oil-formation

History of Waterflooding

Waterflooding was discovered as far back as the mid 19th century and progressively developed since that time[2,3,4]. The first waterflood occurred as a result of accidental water injection in the Pithole City area of Pennsylvania as early as 1865. Many of the early waterfloods occurred by leaks from shallow water sands or by surface water accumulations entering drilled holes. In the late 1800s, the primary function of water injection was to maintain reservoir pressure, permitting wells to have a longer life than allowed by natural depletion.

In the earliest days, water was injected first at a single well, and then as the adjacent wells were watered out, those wells were converted as water injectors, forming a "circle drive." Next, waterflooding developed as "peripheral floods." In another variation, a series of wells were used as injectors, constituting a "line drive." In 1924, the first "five-spot pattern flood" was initiated in the Bradford Field in Pennsylvania.

The application of waterflooding grew from Pennsylvania to Oklahoma in 1931 in the shallow Bartlesville sand, and in 1936 to Texas in Fry Pool of Brown County. In spite of the localized activities in Oklahoma and Texas, waterflooding did not have widespread application until the early 1950s.

Importance of Integrated Waterflood Asset Management

The modern waterflood asset management requires not only technical and operational skills but also business, political, and environmental knowledge. Formulating a comprehensive waterflood asset management plan involves:

1. Depletion and development strategies
2. Data acquisition, analysis and management
3. Geological and geophysical evaluation
4. Reservoir modeling and performance prediction
5. Facilities requirements
6. Economic optimization

Waterflood design, evaluation and management are more effective when all functions, e.g., geology and geophysics, reservoir and production engineering, facilities engineering, drilling, field operations, and related disciplines, work together. Implementing the plan requires management support, field personnel commitment, and multidisciplinary integrated teamwork. Success of the project depends upon careful monitoring/surveillance and thorough ongoing evaluation of its performance, and taking proactive actions in a timely manner.

Integrated waterflood management is essential in creating maximum economic value and oil recovery for the asset.

The State of the Art

Waterflood asset management has advanced tremendously during the past 40-50 years. The techniques and tools are better, improved design and surveillance techniques have come in practice, and automation using local computers has helped data collection, analysis and management.

The synergism and teamwork provided by truly multifunctional teams have been successful. Team members are beginning to work more like a "well-coordinated basketball team" rather than "a relay team." Using an integrated approach to waterflood asset management, along with the latest technological advances will allow companies to maximize economic recovery during the life of the waterflood. Because production in all waterfloods declines over time, innovations that prolong cost-effective waterflood recovery should be of global interest.

Scope and Objective

Important publications on waterflooding include SPE Monograph Volume 3 on "The Reservoir Engineering Aspects of Waterflooding" by F. F. Craig, Jr. in 1971, SPE Textbook Series, Volume 3 on "Waterflooding" by G. P. Willhite in 1986, and SPE Monograph Volume 11 on "The Design Engineering Aspects of Waterflooding" by S. C. Rose, J. F. Buckwalter, and R. J. Woodhall in 1989.[2, 5, 6] The objective of this book is to provide readers with a broader and better understanding of the practical approach to waterflood asset management as a whole entity by integrating the technologies and activities of the many disciplines involved in waterflood projects. This will provide better resource management practices enhancing hydrocarbon recovery and maximizing profitability. The authors have packaged under one cover their knowledge and hands-on experience acquired over many years of their professional careers.

This book is written for the practicing engineers, geoscientists, field operation staffs, managers, government officials, and others who are involved with waterflood asset management. The college students in petroleum engineering, geoscience, economics, and management can also benefit from this book.

Organization

Integrated waterflood asset management involving goal setting, planning, monitoring, evaluating, and revising plans throughout the life of the project is the key to successful operation. This book presents:

- Fundamentals of waterflood asset management
- Technical, operational and economic aspects of waterflood asset management
- Real life examples to show best practices in waterflood asset management

The first chapter discusses the reasons for and history of waterflooding, importance of integrated waterflood asset management, the state of the art, scope and objective, and organization.

Chapter 2 presents fundamentals of waterflood asset management, and waterflood screening criteria.

Integrated technology, involving geology, geostatistics, geophysics, and reservoir engineering, plays an important role in developing reservoir description. Data management, factors influencing waterflood recovery, and infill drilling are important aspects in managing waterfloods. These technical perspectives are presented in chapters 3, 4, 5, and 6.

Chapter 7 presents the design of waterfloods from the points of view of geological, engineering, and operational aspects. An example of waterflood design is also presented.

Waterflood production and reserves forecast, surveillance techniques, field operations, and economics are necessary for effective waterflood asset management, and these are described in chapters 8, 9, 10, and 11.

Chapter 12 presents case histories of selected waterflood asset management with overall analysis including key achievements, inadequacies, and future operating plans.

Chapter 13 addresses current challenges (both technical and organizational), areas of improvement, outlook and the next step.

Details and supporting materials are presented in the appendices.

References

1. Satter, A. and G. C. Thakur. "Integrated Petroleum Reservoir Management: A Team Approach," PennWell Books, Tulsa, OK (1994).
2. Craig, F. F., Jr. "The Reservoir Engineering Aspects of Waterflooding," SPE Monograph Volume 3 (1971).
3. History of Petroleum Engineering, API, Dallas, TX (1961).
4. Fettke, C. R. "Bradford Oil Field, Pennsylvania and New York," Pennsylvania Geological Survey, 4th Series (1938) M-21.
5. Willhite, G. P. "Waterflooding," SPE Textbook Series, Volume 3 (1986).
6. Rose, S. C., J. F. Buckwalter, and R. J. Woodhall. "The Design Engineering Aspects of Waterflooding," SPE Monograph Volume 11 (1989).

2

Reservoir Management Concepts, Process, and Waterflood Prospect Screening

The goal of waterflood asset management is to maximize profits by optimizing recovery while minimizing capital investments and operating expenses. Success in waterflood asset management requires goal setting, deliberate planning, implementing, on-going monitoring, and evaluating the reservoir performance, keeping short term but—more importantly—long term goals on focus.

This chapter provides reservoir management definition and discusses reservoir management concepts and process, synergy and teamwork, and integration of various functions. Also, the waterflood screening and evaluation are covered.

The fundamentals of waterflood asset management, which have been described in detail in Reference 1, are summarized in this chapter. Based upon Reference 2, Appendix A presents a summary of reservoir management definition, integrated reservoir management concepts and process, synergy and team work, organization and team management, and integration of geoscience and engineering. It also discusses reasons for failure of reservoir management program.

Waterflood Asset Management

The function of waterflood management is to provide facts, information, and knowledge necessary to control operations and obtain the maximum possible economic recovery.

Guidelines for waterflood asset management should include information on (1) reservoir characterization, (2) estimation of pay areas containing recoverable oil, (3) analysis of pattern performance, (4) data gathering, (5) well testing and reservoir pressure monitoring, and (6) well information database.

An integrated team approach involving geoscience and engineering professionals, field personnel, and management is essential for waterflood asset management. Successful operations require:

1. Developing an economically viable plan
2. Implementing the plan
3. Monitoring and evaluating performance
4. Revising plans and strategies

These items are discussed in detail in Appendix A.

Waterflood Recovery Efficiency

The overall waterflood recovery efficiency is given by

$$E_{RWF} = E_D \times E_V \qquad (2\text{-}1)$$

Where:

E_{RWF} = Overall waterflood recovery efficiency, fraction

E_D = Displacement efficiency within the volume swept by water, fraction

E_V = Volumetric sweep efficiency, fraction of the reservoir volume actually swept by water

Displacement Efficiency

Displacement efficiency is influenced by rock and fluid properties and throughput (pore volume of water injected). It can be determined by laboratory floods, frontal advance theory, and empirical correlations (See chapter 8 and Appendix D). Factors affecting displacement efficiencies are listed in Table 2-1.

Table 2-1 Factors Affecting Waterflood Efficiencies

Displacing Efficiency

    Oil and Water Viscosities
    Oil Formation Volume Factors at the start and end of flood
    Oil Saturations at the start and end of flood
    Relative Permeability Characteristics

Sweep Efficiencies

    Reservoir Heterogeneity
    (areal and vertical variations in porosity, permeability, and fluid
    Properties)

    Directional Permeability

    Formation Discontinuity/Faults

    Horizontal and Vertical Fractures

    Formation Deep

    Flood Pattern Type

    Cross-Flooding

    Throughput

    Oil/Water Mobility (effective permeability/viscosity) Ratio

---

Sweep Efficiency

Volumetric sweep efficiency is defined by:

$$E_V = E_A \times E_I \tag{2-2}$$

Where:

E_A = Areal sweep efficiency

E_I = Vertical sweep efficiency

Sweep efficiency is related to permeability variations, fluid properties, fluid distribution, fluid saturation and fracture systems. Adverse permeability variations result in poor sweep efficiency, rapid water breakthrough, and high water production. Factors affecting sweep efficiencies are shown in Table 2-1.

Figure 2-1 shows a zone of high permeability at the top of a formation. Water, which follows the path of the least resistance, preferentially enters this zone. The result is early breakthrough and a substantial amount of oil remaining in the reservoir due to poor vertical sweep efficiency.

Since the permeability variation factor is calculated from core analysis, it is vital that core data be collected before starting a waterflood. This data could also be used to indicate zones of high permeability and any major source of excess water production.

Alternately, waterflood recovery efficiency can be calculated as follows:

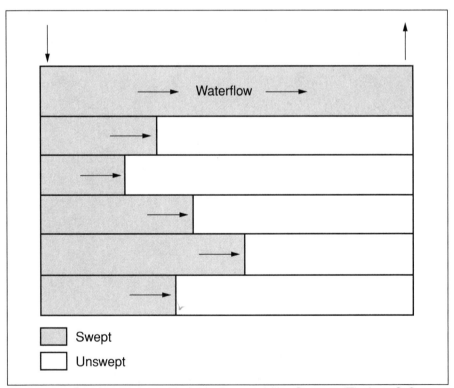

Figure 2-1 Displacement in Variable Permeability System (Thakur, G.C. "Reservoir Management of Mature Fields," IHRDC Video Library for Exploration and Production Specialists, Boston, MA, 1992)

Waterflood Recovery Efficiency = (Original-Oil-In-Place - Primary Recovery - Residual Oil In Conformable (Swept) Area - Remaining Oil In Non-Conformable (Unswept) Area)/ Original-Oil-In-Place. The various parts in this expression are given below:

$$\text{Original Oil In Place (OOIP), bbl/acre-foot}$$
$$= 7,758\phi \text{ x}(S_{oi}/B_{oi}) \tag{2-3}$$

$$\text{Primary Recovery, bbl/acre-foot}$$
$$= 7,758\phi \text{ x}((S_{oi})/B_{oi})\text{x PRF} \tag{2-4}$$

$$\text{Residual Oil/Conformable Area, bbl/acre-foot}$$
$$= 7,758\phi \text{ x } E_v \text{ x}(S_{oc}/B_{oc}) \tag{2-5}$$

$$\text{Remaining Oil/Non-Conformable Area, bbl/acre-foot}$$
$$= 7,758\phi \text{ x } (1\text{-}E_v)\text{x}(S_{onc}/B_{onc}) \tag{2-6}$$

Where:

ϕ = porosity (fraction)
S = saturation (fraction)
B = oil formation volume factor (reservoir barrel/stock tank barrel)
E = sweep efficiency (fraction)
PRF = primary recovery factor (fraction)

Subscripts *i, o, c, nc*, and *v* denote *initial, oil, conformable, non-conformable* and *volumetric*, respectively

Note that $S_o = 1\text{-} S_w$ (subscript w denotes water) for the medium filled with oil and water only. S_{oc} is the residual oil saturation (S_{or}) in the conformable or water flooded area. S_{onc} is the remaining oil saturation in the non-conformable or unswept area and approximately equal to $(1\text{-}S_{wi} -S_g)$, where S_g is the free gas saturation after primary depletion. Since the pressure at the conformable and non-conformable areas could be virtually the same, B_{oc} and B_{onc} can be considered to be the same as B_{or}, which is approximately equal to 1.

Using Equations (2-3) through (2-6), waterflood recovery efficiency is given by

$$E_{RWF}, \text{ fraction OOIP} = ((S_{wi}/B_{oi})\text{x}(1\text{-PRF})\text{- } E_v \text{ x } (S_{or}/B_{or})$$
$$-(1\text{-}E_v)\text{x}((1\text{-}S_{wi}\text{-}S_{gi})/B_{or}))/((1\text{-}S_{wi})/B_{oi}) \tag{2-7}$$

Then, the waterflood recovery efficiency is dependent upon five primary variables: S_{wi}, S_{or}, B_{oi}, PRF, and E_v. Considerable latitude can be tolerated in any one of the factors, provided the other factors are favorable. A combination of unfavorable conditions for two or more of the factors will result in a poor waterflood. If all the primary variables are favorable, we can hope for a high waterflood recovery. In addition, there is still a reasonable chance of successful waterflood if only the initial or connate water saturation and the sweep efficiency are favorable. Thus, these two variables are of paramount importance and need to be evaluated most critically.

Primary Recovery Factor

Primary recovery factor due to solution gas drive is influenced by porosity, initial water saturation, oil formation volume factor at the initial or bubble point pressure, absolute permeability, oil viscosity at the initial or bubble point pressure, bubble point and abandonment pressures 3 . There is usually no major problem in calculating primary recovery efficiency for fields that have been recently developed and have adequate reservoir and production data. However, it is difficult to estimate primary recovery factors in old fields that have unreliable production data or have been subjected to unsatisfactory logging methods.

Primary and other variables affecting waterflood recovery are listed in Table 2-2.

Waterflood Prospect Screening

Some "Rules of Thumb" of screening criteria for quick evaluation are given in Table 2-3, including empirical performance factors for injection and flood response.

Fluid-Saturation, Distribution and Properties

One of the most important factors responsible for the success or failure of a waterflood is the fluid saturation at the start of the flood. It is difficult to have an economic waterflood at water saturation of 50 % or more. Under these high saturation conditions, the relative permeability of water has an adverse effect, making it difficult to form an oil bank.

Table 2-2 Factors Affecting Waterflood Recovery

Primary Variables	Other Variables
Initial Water Saturation	Porosity
Residual Oil Saturation (Displacement Efficiency)	Absolute and Relative Permeability Characteristics
Oil Formation Volume Factor (Crude Shrinkage)	Oil/Water Viscosity
Primary Recovery Efficiency	Initial Pressure
Volumetric Sweep Efficiency	Abandonment Pressure After Primary Depletion
Areal Sweep	Structural Characteristics
Vertical Sweep	Reservoir Heterogeneity
	Flood Pattern
	Time of Flood
	Economic Factors
	Well Spacing
	Reservoir Depth
	Oil Price
	Water Availability
	Operating Costs

However, it is possible to conduct a successful waterflood with connate water saturations as high as 50% or more, only if a favorable combination of other factors is present. In general, higher connate water saturations increase the risk involved in waterflooding.

The relative permeability characteristics of a reservoir rock are a measure of the rock's ability to conduct one fluid when one or more fluids are present. These flow properties are based on pore geometry, wettability, fluid distribution and saturation history. Figures 2-2 and 2-3 show typical water/oil relative permeability characteristics for water-wet and oil-wet formations (See more in chapter 4).

The water/oil flow properties of reservoir rock are generally used to estimate the oil recovery we might obtain by flushing an oil-saturated rock with water. Information on the relative permeability to water at floodout conditions provides the water injectivity value.

Table 2-3 Rules of Thumb-Screening Criteria for Quick Evaluations

Empirical Performance Factors for:

	Range of Values	Average Design Factor	Comments
❏ Injection Rate – Pattern Flood – Aquifer injection	2 to 5 BWPD (Barrels of Water Per Day)/net ft, 5 to 15 BWPD/net ft		Estimate from Actual Injectivity Tests, Empirical Charts, or Local Experience
❏ Ultimate Water Injection	0.9 to 2.0 PV (Pore Volume)	1.5	
❏ Ratio of Ultimate Water Inj. to Ultimate Incremental Oil Prod.	5.6 to 70.8	15	

(left margin label: Injection)

Empirical Performance Factors for:

	Range of Values	Average Design Factor	Comments
❏ Flood Response – Time to Kick – Fill-up at Kick	2/3 of Fill-up 0 to 45 Months 0.20 to 0.67 PV	0.5 PV	Use a Lower Value for Heterogeneous Layered Reservoirs
❏ Time to Reach Peak Oil – Fill-up at Peak	0.54 to 3.24 PV	1 PV	At the Time of Fill-up, the Peak Rate Is Achieved Quicker but Is Lower for a Heterogeneous, Layered Reservoir
❏ Length of Peak	6 to 49 Months		Length of Peak Will Be Longer if Injectivity is Poor
❏ Ratio of Cumulative Waterflood Oil at End of Peak to Ultimate Waterflood Oil	0.33 to 0.63	0.45	
❏ Ratio of Peak Oil Rate to Injection Rate	0.03 to 0.55	0.2	
❏ Oil Recovery Before Peak	1/2 of the Remaining Primary and Secondary Recovery		
❏ Oil Decline Rate After Peak	10% to 25% per Year		
❏ Total Production Rate	80% of Water Injection Rate (Injection Efficiency)		
❏ Secondary to Primary Ratio	1/2 to 1		Depends on Permeability, Dykstra-Parsons Variation Factor, Oil-Water Viscosity Ratio, and Well Spacing
❏ Ultimate Water Production	1/2 to 3/4 of Ultimate Water Injection		Losses Due to Gas Cap Fill-up and Out-of-Zone Injection
❏ Minimum Oil Saturation at the Start of Waterflood	50% (40% Minimum)		

(left margin label: Flood Response)

Guidelines

❏ **What You Need to Do:** Calculate Oil Saturation at Start of Waterflood (S_o), % pore volume

❏ **How to Do It:**

$$S_o = (1 - S_w) \left(1 - \frac{N_p}{N}\right)\left(\frac{B_o}{B_{oi}}\right)$$

Where

S_w = Water Saturation at Start of Waterflood, Fraction of Pore Volume

N_p/N = Cumulative Oil Production at Start of Waterflood, Fraction of STOIP

B_o = Oil Formation Volume Factor at Start of Waterflood, res bbl/STB

B_{oi} = Initial Oil Formation Volume Factor, res bbl/STB

❏ **Things to Remember:**
– Check the Reservoir to See if All Areas Have Equal Drainage. This Check Ensures the Flooded Area's Oil Saturation will Not Be Lower than the Average for the Reservoir
– Any Area with <40% Oil Saturation is a Poor Flooding Prospect

❏ **Range of Values for S_o:** 40-67

Another important factor to consider in waterflood screening is the gas saturation at the start of the flood. The gas, which is liberated in the reservoir by pressure depletion, must be resaturated. If low sweep or displacement efficiencies occur during a waterflood, a large proportion of the oil may resaturate portions of the unswept pore volume and not

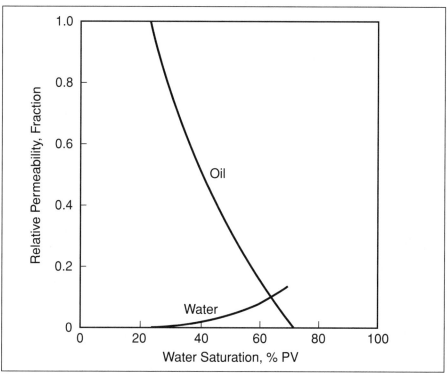

Figure 2-2 Typical Water/Oil Relative Permeability Characteristics, Strongly Water-wet Rock (Craig, Jr., SPE Monograph 3, *"The Reservoir Engineering Aspects of Waterflooding,"* Richardson, TX, 1971)

be produced. Reservoirs with high formation volume factors and a high solution gas/oil ratio should be flooded at, or slightly above, the bubble point where the viscosity is the minimum.

In thick reservoirs with high horizontal and vertical permeability, fluids may be unevenly distributed. In these cases, the top of the reservoir will contain a high gas saturation, and the base of the sand will contain a high oil saturation. Water injected in this type of reservoir will tend to enter the formation having high gas saturation and will override (and often bypass) the oil. Similar problems may occur in reservoirs that are underlain by water. In these cases, the water may have a tendency to under run the oil.

The properties of the reservoir fluid (Figure 2-4) and the injection fluid affect the flood performance. In general, reservoirs containing viscous oil perform relatively poorly in response to a waterflood (See

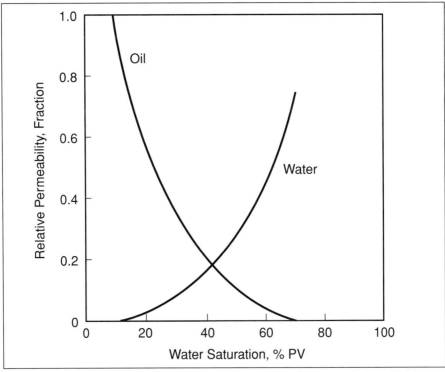

Figure 2-3 Typical Water/Oil Relative Permeability Characteristics, (Craig, Jr., SPE Monograph 3, "*The Reservoir Engineering Aspects of Waterflooding*," Richardson, TX, 1971)

chapter 5). When water displaces oil less mobile than itself, the displacement front is unstable, and the water has a tendency to "finger" through the reservoir, causing excessive water production.

Fractures

If injection and producing wells are located along a line parallel to fracture directions, early breakthroughs occur (Figure 2-5). If injection wells are located along a line parallel to the direction of the fracture, interference between injectors will cause water to move in a direct line across the fractures towards producing wells. Thus, it is important to know the fracture direction before designing a waterflood pattern. This one factor may dictate the success or failure of a waterflood.

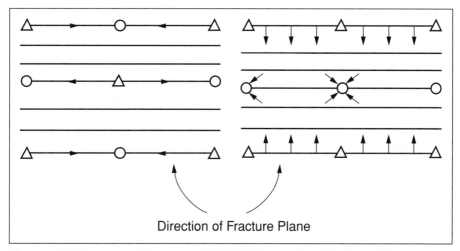

Figure 2-4 Schematic Diagram of Oil PVT Properties (Thakur, G.C.
"*Reservoir Management of Mature Fields*," IHRDC Video Library for
Exploration and Production Specialists, Boston, MA, 1992)

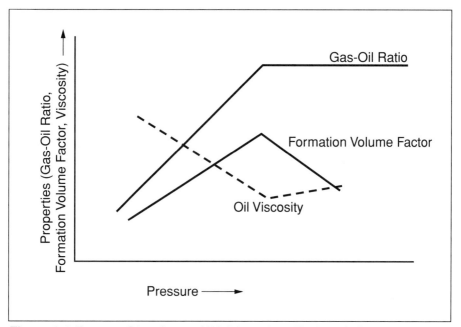

Figure 2-5 Fracture Direction and Well Location (Thakur, G.C. "*Reservoir
Management of Mature Fields*," IHRDC Video Library for Exploration and
Production Specialists, Boston, MA, 1992)

Areal and Vertical Reservoir Heterogeneity

Areas of high and low permeability in a reservoir may cause an unbalanced flood performance. Cross bedding may also impair fluid movement between injection and production wells. Sometimes, a reservoir may contain planes of weaknesses or closed natural fractures that open at bottom hole injection pressures.

In addition to these factors, we must also consider the level of reservoir continuity between an injection/production well pair.

Reservoir Pressure Level

Waterflooding results in maximum oil recovery when the reservoir pressure level is at the original bubble point (See chapter 5). At the original bubble point pressure, a barrel of stock tank oil represents the maximum amount of reservoir oil, thereby occupying the largest volume and consequently a high oil relative permeability. In addition, several other factors favor waterflooding at this pressure:

- Reservoir oil viscosity is at its minimum value, which improves the mobility and areal sweep
- Producing wells are at their highest productivity index

There is no delay in flood response, since the reservoir has no free gas saturation at this point.

Some people argue that the optimum pressure level is slightly below the bubble point pressure because the presence of a small amount of gas saturation reduces the residual oil saturation to waterflood. However, it is difficult to practically achieve this condition uniformly throughout the reservoir. A large variation of gas saturation may occur and negatively affect the waterflood performance.

References

1. Thakur, G. C. "Waterflood Surveillance Techniques – A Reservoir Management Approach," *JPT* (October 1991): 1180-88.
2. Satter, A. and G. C. Thakur. *Integrated Petroleum Reservoir Management: A Team Approach*, PennWell Books, Tulsa, Oklahoma (1994).
3. Arps, J. J., et. al. "A Statistical Study of Recovery Efficiency," API Bulletin D14 (1967): 1-33.
4. Craig, F. F., Jr. "The Reservoir Engineering Aspects of Waterflooding," SPE Monograph 3, Richardson, Texas (1971).
5. Thakur, G. C. "Reservoir Management of Mature Fields," IHRDC Video Library for Exploration and Production Specialists, Boston, MA (1992).

3

Intregrated Technology —
Geoscience and Engineering

This chapter emphasizes the importance of geoscience in reservoir characterization influencing waterflood asset management and discusses methodologies for developing reservoir models. The details of the reservoir modeling are described in Appendix B.[1] It is important for the waterflood asset management team to understand the concepts used to develop a reservoir model and be aware of the limitations that are present in the process. In addition, several case studies illustrating the importance of reservoir description are presented in this chapter. The examples of waterflood experience emphasizing reservoir description process illustrate the importance of developing a sound reservoir model for waterflood asset management.

The purpose of a well-integrated study is to assist in answering the following questions:

1. Will the reservoir perform:
   - as a series of independent layers?
   - as zones of different permeability with crossflow?
2. Are there zones with high gas or water saturation that could be channels?
3. Does the reservoir contain long natural fractures that could cause directional permeability and preferential fluid movements?
4. Are there areas of high or low permeability that may cause unbalanced flooding?
5. Will cross bedding impair communication between injectors and procedures?
6. Does the reservoir contain closed fractures that could open at bottom-hole injection pressures?

Organizing and Collecting Data

Recording, interpreting, and applying reservoir data correctly are some of the most important tasks that petroleum engineers and geologists perform. The quality and quantity of these data determine the success of any subsequent geological and engineering studies and, consequently, of the project as a whole.

Because data collection and handling are so important to project success, they must be carefully planned and carried out. This entails formulating a systematic data collecting program based on the following criteria:

- We must first have a clear understanding of the data's purpose and application—that is, we must be able to explain *why* we need the information and *what* we are going to do with it before proceeding with data collection
- We must obtain the most complete coverage of data and testing possible among the reservoirs and wells, given the funds available for data collection
- We must develop and use a consistent procedure to ensure that the data collected represent actual reservoir conditions and can be compared to each other
- We must make the most effective and comprehensive use of the data

Table 3-1 lists sources for various types of reservoir data, and Table 3-2 describes various measurements and observations of data. These data may be divided into two groups: *static* and *dynamic*. *Static* data represent direct measurements of some fixed property of the reservoir or its fluids (for example, the porosity, permeability, connate water saturation, temperature, chemical composition, etc.). *Dynamic* data refer to the level of forces contained or induced into the reservoir (for example, pressure, PVT properties, effective permeability obtained from transient testing, skin factor, and flood fronts).

Porosity and permeability are necessary properties for defining net pay. We measure these properties primarily through core and log analysis.

Reservoirs rarely are completely cored. Since more log data are available than core data, logs are usually correlated with core porosities if both are available, and then core equivalent porosities are derived from the log data (Figure 3-1). If data points are widely scattered, the geologist should determine if these variations are caused by lumping together data from different lithologies.

Table 3-1 Data Sources (Courtesy Petroleum Engineer International, May 1980)

Time	Predrilling							During Dilling													Post Development					Special Studies
Operation	Gravity	Seismic			Geology-Eng. Study		Well Bore Operations					Logs						Wire Line		Production						
	Gravity	Time	Velocity	Amplitude	Character	Analogy, Regional Knowledge and Maps	Depositional Environment	Drill Rate	Mud Log	Cuttings	Cores	Drillstem	Electric	SP	Acoustic	Density	Gamma Ray	Neutron	Test	Cores	Flow Test	Pressure	Water Cut	GOR	History	Analogy	Engineering and Geology
Depth Makers	2	2				2	2	3	3	2	1	1	1	1	1	1	1	1	2	1	2						1
Structure and Area	2	2	1	3	3	2							4								2	2	3	3	1		1
Hydrodynamics						1															2	1		3	1		1
Gross Thickness			2		3	2	2	2	2	3	2	4	1	1	1	1	1	1			2						1
Net Thickness				2	2	2	2	3	3	4	1	4	1	1	1	1	1				2						1
Lithology			2	2	3	2	2	3		2	1		3	3	2	2	2	3	1							2	1
Mechanical Properties			2	2	3	2	2	3		2	1				2	2	2	3		2						2	1
Contacts			2	2	2	4		3	2	2	2	1	2	1	1		1	1	2		1		2	2	2		1
Pressure			2	3		1		3				1								2	1	1			1		1
Porosity			2	2	3	2	2	4		3	1	3			1	1		1		2	4					2	1
Permeability					4	2	2			4	1	1	4		3	3		3	2	3	2	1				2	1
Relative Premeability										1											1	2	2	2	2	2	1
Fluid Saturation			3	3	3	4		3	3	2	1	1	3	2	2	2	2		3		1		1	1	2		1
Pore Sizes						2				2	1	4	4	4	4	4	4	4	3							2	
Producing Mechanisim	4		3	3	3	2	3																1	1	1	1	1
Hydrocarbon Properties			4	4		2		3	4	3	1					4	4	4	2	4	1	2		1		2	1
Water Properties						1				4	1								2		1		1			2	1
Production Rate						2	2			2	2					4	4	4	2	3	1	1			1	2	1
Fluids Produced										1											1		1	1	1	2	1
Well Damage																					1	1			1		1
Recovery Efficiency																							2	2	1	2	1

Code: 1. Best Source, 2. Good Data Source, 3. Average Data Source, 4. Poor Data Source
Also, See World Oil, November 1978, p.57

Table 3-2 Measurements and Observations of Data

Parameters	Measurements and Observations
• Rock types - macroscopic and microscopic	Visual thin section, SEM (scanning electron microscope), empirical log response, seismic
• Paleontologic and organic constituents	Visual
• Porosity	Visual core analysis, log calculations, mass balance, seismic
• Permeability	As with porosity plus pressure and flow rate data
• Fluid types	Sample analysis, log response, chromatograph, drilling fluid changes, seismic
• Sample fluorescence	Visual
• Bed thickness	Log response, drill time changes, lithologic descriptions, seismic
• Structural relationships to other wells	Correlation, seismic
• Stratigraphic relationships with other wells	Correlation, seismic
• Bedding dip (structural and stratigraphic)	Visual, dipmeter calculations, seismic
• Drilling penetration rates	Drill time changes, well history
• Formation damage	Production rates, pressure analysis
• Drilling problems	Well history and drill rate, seismic
• Formation pressure gradients	Pressure analysis, seismic
• Drilling fluid gradients	Drilling fluid history
• Fracture gradients	Calculate from formation, density formation pressure, and drilling fluid pressure
• Hydrodynamic gradients	Pressure analysis

Logging devices (including sonic, formation density and neutron logs) measure the average porosity over the interval of investigation. Porosity obtained from these logs is a good index for distinguishing between pay and non-pay zones when used in conjunction with data obtained from the microlog. Micrologs read point-by-point data and can also indicate the presence of a filter cake buildup and, therefore, suggest a permeability. In mature reservoirs, old logs are often available. These old logs must be calibrated with new logs and/or core data, and then used to define reservoir properties where no new data are available.

For most mature reservoirs, the permeability measurement used in routine analysis is usually an air value, which must be corrected for the

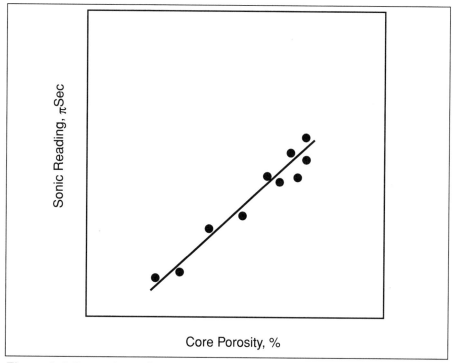

Core Porosity, %

Figure 3-1 Crossplot of Sonic Reading and Core Porosity (Thakur, G.C. "Reservoir Management of Mature Fields," IHRDC Video Library for Exploration and Production Specialists, Boston, MA, 1992)

Klinkenberg effect (see Figure 3-2). The permeability of a reservoir may often be considered to consist of two systems: a matrix and a vug (or fracture) system. The vertical permeability is generally small for stratified beds, but it may be essentially the same as the horizontal permeability for non-stratified, homogeneous reservoirs.

In the case of an old reservoir where only a few core plugs were evaluated, we will often find no correlation between the results of a flow test and the porosity and/or permeability of the cores analyzed.

Facies maps are created by defining various lithological characters. A more detailed description of the facies mapping is available later in this chapter.

The *connate water saturation* of a given volume of rock is related to its structural position, its pore geometry and its wettability. Connate water content often shows a rough relation to permeability, with lower-permeability rocks having more connate water. Also, formations with a higher clay and silt content generally have higher connate water saturation.

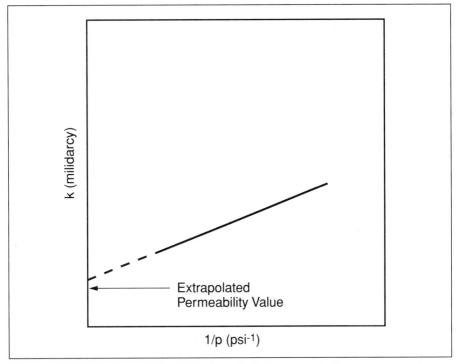

Figure 3-2 Klinkenberg Effect (Thakur, G.C. "Reservoir Management of Mature Fields," IHRDC Video Library for Exploration and Production Specialists, Boston, MA, 1992)

The method of choice for determining connate water saturation is by analysis of cores cut with an oil-base coring fluid. Make sure that the coring fluid does not affect the rock wettability. Most old reservoirs either lack these data or have only unrepresentative data available due to drilling and retrieval and storage practices.

The next best method for estimating connate water content is the *capillary pressure*, or *restored-state method*. Reservoir engineers and geologists often compare connate water saturation as determined by capillary pressure with an oil-base core, where available. If the data compare satisfactorily, the capillary pressure method can be used. Note that the air-mercury method should be used cautiously for the estimation of connate water saturation.

Figure 3-3 illustrates the types of reservoir and well data needed for a reservoir study. Tables 3-2 through 3-5 list the major sources of these data. Tables 3-6 and 3-7 outline typical data base requirements, while Figure 3-4 outlines a comprehensive plan for reservoir data processing.

Table 3-3 Formation Evaluation Data (Thakur, G.C. "Reservoir Management of Mature Fields," IHRDC Video Library for Exploration and Production Specialists, Boston, MA, 1992)

In-situ Measurements	Surface Measurements	Geological Interpretation
Mud logging		
Sample Logs	Core analysis	Log normalization
Oil, gas shows	Porosity	Gross/net thickness
Cuttings analysis	Permeability	Lithology maps
Drill stem	Relative permeability	GOC/WOC determination
Permeability	Wettability	Cross sections
Skin	Fluid distribution	Permeability maps
Fluid samples	(connate water)	Clay identification
Well logging	Capillary pressure	Structure
Electrical	Pore geometry	Gross/net isopach maps
• Spontaneous	Rock compressibility	Reservoir volumes
Potential (SP)	Sidewall core properties	
• Non-focused electric		
• Focused resistivity/ conductivity		
• Induction		
• Focused and nonfocused micro resistivity		
• Dielectric		
Acoustical		
• Sonic		
• Cement bond		
Radiation		
• Formation density		
• Gamma ray		
• Neutron-gamma ray		
• Neutron-thermal		
• Neutron-capture		
• Carbon/oxygen		
Miscellaneous		
• Dipmeter		
• Repeat formation tester (RFT)		
• Mud log/drill cutting		

Table 3-4 Reservoir Fluid and Production Data (Thakur, G.C. "Reservoir Management of Mature Fields," IHRDC Video Library for Exploration and Production Specialists, Boston, MA, 1992)

Fluid analysis (oil, gas, water)	Pressure data	Production-injection data	Completion-workover records	Chemical program
• Fluid composition • PVT behavior • Com-pressibility • Viscosity • Specific gravity • Solubility • Chlorides (injection-formation water)	• Original completion data • Annual BHP survey • Build-ups • Drawdown • Injection falloffs • Fracture pressure (step-rate tests) • Static BHP tests • Inter-ference tests • Hall plots • Pressure monitoring program	• Yearly/monthly fluid volumes (O/W/G) • Allocation factor • Fluid input/output ratios • Gas/oil ratios (annual surveys, production tests, etc) • Tracer surveys • Docu-mentation of known channels • Periodic testing program	• Well file reviews • Completion similarities • Cementing procedures • Perforating procedures • Stimulation histories • Squeeze procedures • Formation damage	• Scale identifi-cation prevention • Paraffin control • Emulsions • Corrosion control

Table 3-5 Surface Facilities Data (Thakur, G.C. "Reservoir Management of Mature Fields," IHRDC Video Library for Exploration and Production Specialists, Boston, MA, 1992)

Production	Injection
Artificial lift	Plant capacity
Beam pumping (unit capacity): SPM (strokes per minute), SL: (stroke length), tubing size, pump size, cycle	Maximum operating pressure
Hydraulic and centrifugal pumping	
Gas lift:	
Tubing and casing pressure	
Separation facility	
Metering accuracy	

Table 3-6 Computer Data. The Computer Database Is Described in Detail in Table 3-7 (Thakur, G.C. "Reservoir Management of Mature Fields," IHRDC Video Library for Exploration and Production Specialists, Boston, MA, 1992)

- Data base
- Location of data set
- Top of zones
- Pattern allocation factor
- Reservoir pressure data set
- Documentation of programs

Table 3-7 Database (Thakur, G.C. "Reservoir Management of Mature Fields," IHRDC Video Library for Exploration and Production Specialists, Boston, MA, 1992)

Database Contents	Pattern Analysis	Voidage	Zonal Allocation	Time Normalization	Incremental Production
• Typical Pattern Data Bulk Volume Areal Allocation Factor HCPV Area FVF GOR Sw Gas/Oil Fuel Use ROS Porosity Zone Thickness Enthalpy • Typical Event Data Zone ID Major/Minor Category Code Completion ID & Number Zone Allocation Factor Workover Evaluation Code • Produced Volumes Oil, Gas, Water, CO2, N2 • Injection Volumes Gas, Water, CO2, N2, Steam • Well Days for Each Volume Type • Field, Reservoir, Property Codes • Well Number & Unique Identifier • Well Performance/IPR Curves	• Computes Production & Injection By Pattern • Uses Areal Allocation Factors • Uses PVT & Petrophysical Data • Reports Results for user Specified Patterns by Group & Sum of Groups • Provides Data Files for Other User Processing	• Determines Reservoir Fluid Voidages • Uses PVT & Petrophysical Data • Calculates Rates & Cumulatives for Production & Injection • Reports Results by Group & Sum of Groups • Provides Data Files for Other User Processing	• Allocates Injection Fluids for All Zones in a Well • Calculates Reservoir Volumes, Rates & Cumulatives • Uses Profile Survey Information, PVT, & Petrophysical Data • Includes Steam Injection Reporting • Reports Results by Zone & Well • Provides Data Files for Other User Processing	• Determines Incremental Production Due to Well Events • Helps Engineers Analyze Cumulative Effects of Well Workover • Adjusts All Event Dates to a Common "Normalized" Time • Uses Decline Program • Reports Results by Well & Group • Provides High Resolution Graphics & Data Files	• Determines Incremental Production Due to Well Events • Presents Results on Real Time Basis • Performs Event or Breakaway Analysis • Uses Decline Program • Reports Results by Well, Group & Grand Total • Provides High Resolution Graphics & Data Files

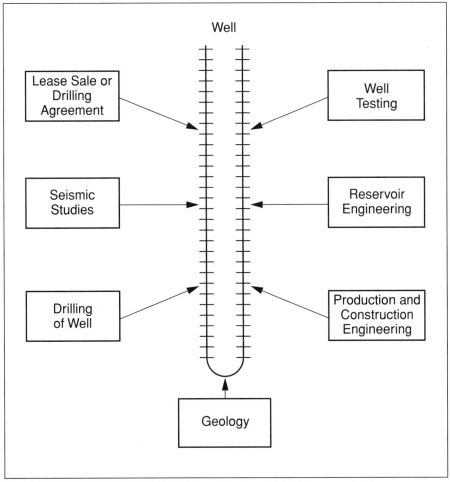

Figure 3-3 Reservoir and Well Data (Thakur, G.C. "Reservoir Management of Mature Fields," IHRDC Video Library for Exploration and Production Specialists, Boston, MA, 1992)

Geological Input

Most reservoirs are not homogeneous, but instead exhibit complex variations in continuity, thickness patterns and other properties, including porosity, permeability and capillary pressure. The reservoir is commonly subdivided vertically and areally into zones or areas based upon differences in rock properties. The complexity of reservoir rock characteristics provides a challenge to earth scientists and engineers, requiring

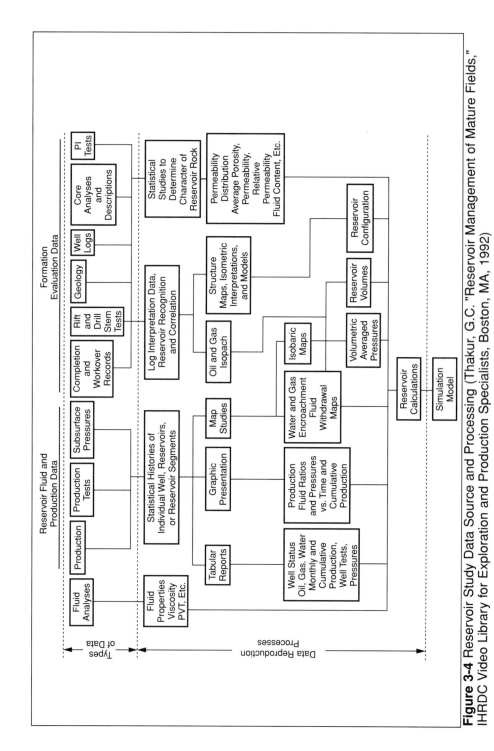

Figure 3-4 Reservoir Study Data Source and Processing (Thakur, G.C. "Reservoir Management of Mature Fields," IHRDC Video Library for Exploration and Production Specialists, Boston, MA, 1992)

them to apply the technologies and experience available in order to improve oil and gas recovery.

Over the years (and after stumbling many times), earth scientists and engineers understand the value of synergism between their functions. In the last 15 years, the emphasis has been on the value of detailed reservoir description, using geological, geophysical and engineering concepts. Earth scientists have been challenged to provide more accurate reservoir descriptions for use in engineering calculations. The nature and degree of reservoir inhomogeneity caused by complex variations of reservoir continuity, thickness patterns and other reservoir properties must be clearly understood before we can develop accurate performance projections or monitor ongoing secondary or enhanced recovery operations successfully. The objective of sound reservoir management can not be reached without identifying and defining all of the individual reservoirs in a given field along with their physical properties.

The best way to identify and quantify rock framework and porespace variations is through the deliberate and integrated use of engineering and earth-science technology. Reservoir studies are more effective when geologists and engineers determine jointly, *at the outset*,

- the course of investigation
- the work-area responsibility for each professional on the description team
- the target dates for combining results

This approach requires an understanding of the technology used by other professionals and an awareness of the principles and concepts upon which that technology is based. This understanding and awareness promotes the free exchange of ideas that forms the basis of successful synergistic activities.

The Geologist's Mapping Process

The purpose of geological mapping is to find traps that contain oil and gas, and once they are found, to apply geologic evidence and concepts toward achieving the most efficient development and production of these prospects. However, it's important to remember that these geologic maps are never finished. When new wells are drilled or old wells are re-examined, new information becomes available, and the maps are changed. Original maps may be based upon a few scattered control points. This means that in the early stages of geological work, a careful study of the local area should be made.

Well data are used in preparing a variety of geological maps, which commonly include:

- structural maps and sections
- isopach maps
- facies maps
- paleogeologic maps

These maps vary widely in scale and amount of detail, depending upon the amount of information available and the purpose for which they are constructed.

Structural Maps and Sections

These displays may be mapped on any formation boundary, unconformity or producing formation that can be identified and correlated by well logs. Structure may be represented by contour elevation maps or by cross sections. A structural cross-section of the Slaughter and Levelland fields is shown in Figure 3-5, and Figure 3-6 shows a structure map on the top of the Mission Canyon formation.

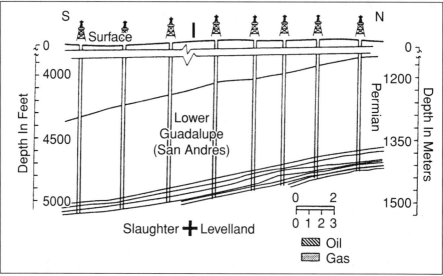

Figure 3-5 Structural Cross Section of the Slaughter and Levelland Fields, Cockran and Hockley Counties, Texas (Thakur, G.C. "Reservoir Management of Mature Fields," IHRDC Video Library for Exploration and Production Specialists, Boston, MA, 1992)

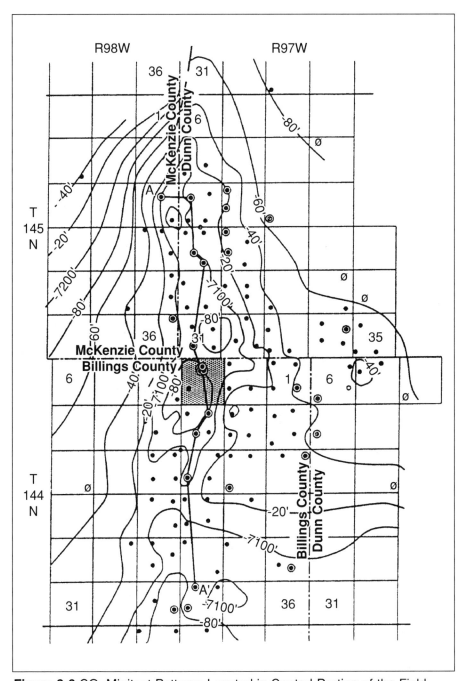

Figure 3-6 CO_2 Minitest Patterns Located in Central Portion of the Field Within Shaded Section

Isopach Maps

These maps use contours to show the varying thickness of the rocks intervening between two reference planes. Isopach maps offer a simple method of showing the distribution of a geological unit in three dimensions. The thickness of individual formations may also be mapped in this manner. Isopach maps may be prepared in minute detail for a particular area, or they may be made for regional studies. A simple example of an isopach map is shown in Figure 3-7.

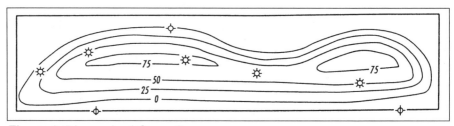

Figure 3-7 Isopach of Gross Pay

Isopach maps are especially useful in determining the time of faulting and folding. They are also useful in regional studies where a geologic history is desired. Often, depositional features (such as sand pinch-outs, sand bars, reefs, and lenses) are related to particular thickness contour patterns shown on the isopach maps. Minor changes in formation thickness become significant displays on an isopach map.

Facies Maps

There are different types of facies maps used in exploration, but the most commonly used are lithofacies maps, which distinguish the various lithologic types. Nearly every formation or group of formations lies within definite stratigraphic boundaries, but within these boundaries one rock type may grade laterally into another. Several devices have been developed to represent changes in rock facies throughout a region on maps. One of the simplest ways is to draw a circle around each well location, and represent the composition percentages of the various rocks in the producing formation by slices, as a pie chart. An example of this type of map is shown in Figure 3-8. The diameter of each circle is proportional to the thickness of the formation mapped in the well. A glance at the map immediately shows general changes in thickness as well as changes in character and relative amounts of the different predominant rock types. Another type of facies map is shown in Figure 3-9,

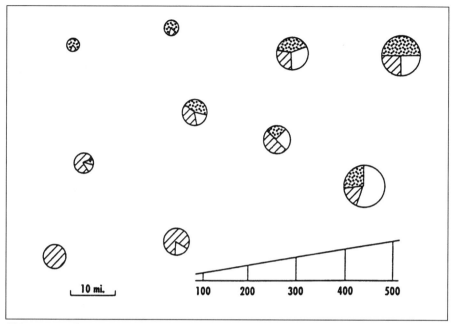

Figure 3-8 Simple Lithofacies Map Showing Only the Evidence of Facies Changes in a Formation or Group of Formations. The Diameter of a Circle Represents the Thickness of the Interval Mapped, The Pie Slices the Percentage of Each Kind of Rock Encountered

which shows the variations in proportion of one particular facies of interest (here, coarse clastics) within a region.

Lithofacies maps are quite helpful in defining various reservoir rock types. The correlation of a number of reservoir properties, including porosity and permeability, becomes much more meaningful when applied to a specific rock type.

Paleogeologic Maps

These are maps that show the paleogeology of an ancient surface. These maps represent an area's geology as it was during various geological periods in the past. They require us to imagine ourselves at some point in the geological past and to consider the geology of the particular area as it existed then.

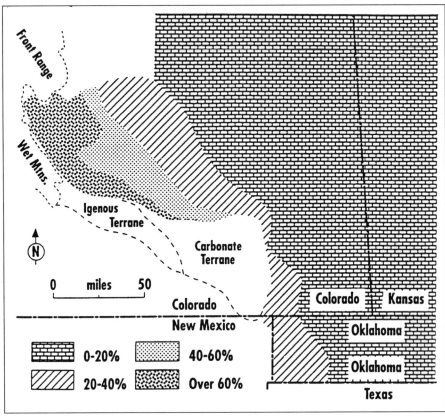

Figure 3-9 Map Showing the Distribution of Coarse Clastics in the Pennsylvanian Rocks of Southeastern Colorado. This Facies Map Shows the Coarsening of the Sediments Toward the West, Presumable Toward the Land Mass That Furnished the Sediments

Reevaluation of Subsurface Data

As more wells are drilled, geological interpretations and maps change. Thus, the geologist is constantly reevaluating subsurface data. Well logs and cores are often used to identify and correlate reservoir rocks and to determine the porosity of potential reservoir rocks and the nature of the fluids they contain.

Common types of well logs include

- drillers' logs
- sample logs (lithological and paleontological)

- electrical logs
- gamma-ray and neutron logs
- drilling time logs
- core and mud analyses
- caliper logs
- temperature logs
- dipmeters

The use of old electric logs has become a most effective and widely used geological tool. The logs are generally used to identify and estimate porosity and reservoir fluid type. Electric logs commonly used include the SP log, resistivity log, microlog, laterolog, microlaterolog and induction log. These logs may date from as far back as 1950-60, but they are still quite valuable in terms of correlating zones or showing the differences between permeable and impermeable formations.

In some reservoirs, the net productive formation, or *net pay*, can be identified by a porosity distribution. A *cut off value* of porosity is selected so that only samples with porosities greater than the cutoff are defined as net pay.

Figures 3-10 and 3-11 show plots of the cumulative volume capacity for a set of data. About 98% of the storage capacity is represented

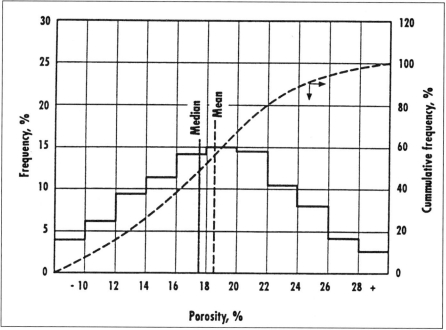

Figure 3-10 Porosity Histogram and Distribution for All Samples From Field A

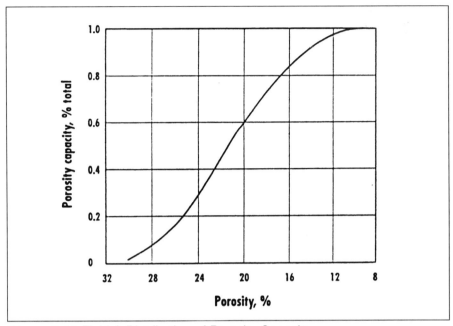

Figure 3-11 Field A Distribution of Porosity Capacity

by samples having porosities of 10% or greater. The 95% porosity capacity corresponds to a porosity of 13% (that is, 95% of the storage capacity is for the samples with porosities of 13% or greater). Thus, if a cutoff value of 13% is chosen, then 95% of the producible hydrocarbons will still be considered.

Unlike porosity, permeability is not normally distributed. Rather, it is "log normally" distributed, i.e., when the logarithm of permeability and frequency are plotted, it yields a normal distribution. Permeability, like porosity, is used to determine the net pay to be used in volumetric calculations. A cutoff value of permeability can be selected on the basis of the samples which have a permeability equal to or greater than the cutoff value. An example of the cumulative permeability capacity versus permeability is shown in Figure 3-12. Eighty % of the producing capacity of this reservoir is represented by samples having a permeability greater than 45 md. Ninety-five % of the capacity is represented by samples having permeabilities greater than 10 md.

Note that the concept of porosity and permeability cutoffs is more appropriately applied on the basis of individual facies. Also, the correlation of porosity and permeability is more meaningful if it is performed for individual facies rather than for the entire field.

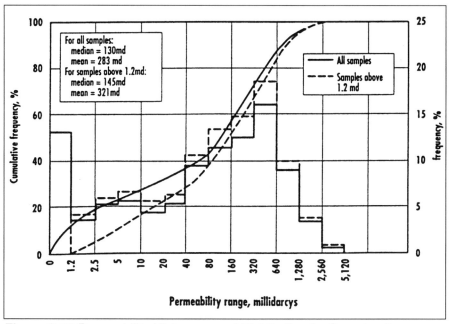

Figure 3-12 Permeability Histogram and Distribution for Samples From Field A

Generating New Maps From Old Data

The volume of rock containing hydrocarbons can be determined for gross or net sections. The rock volumes of the gross section can be determined from the contour maps or the isopach maps. The net rock volume is determined from the net isopach map.

The areas enclosed by contours of the structure maps, based on the top and bottom of the formation, are determined with a *planimeter* (a measuring instrument that calculates the area of an irregular figure by tracing its perimeter). These areas are plotted as a function of depth. The area enclosed by the two resulting curves represents the gross volume of hydrocarbon bearing rock, which is determined by graphical integration or by planimetering. (It is common practice these days to perform this task on a computer. Programs are available to digitize maps, automatically set up a grid system and calculate the rock volume.)

Various methods can be used to calculate the rock volume from planimeter data of an isopach map. The volume can be calculated by the trapezoidal rule:

$$Volume = h/2 \; (A_o + 2A_1 + 2A_2 + \cdots + 2A_{n-1} + A_n) + h_n \, A_n$$

where

h = isopach contour interval
A_o = area enclosed by the zero thickness contour
A_i = area enclosed by each successive (i = 1, 2, 3, - - -) contour
A_n = area enclosed by the contour line representing the greatest thickness
h_n = average thickness above the top contour

The pyramidal rule can also be used to calculate the net volume. The volume of space occupied by hydrocarbons is defined by:

$$V_h = V_b \; \phi \; (1 - S_w)$$

where

V_h = reservoir hydrocarbon volume
V_b = bulk rock volume containing hydrocarbons
ϕ = mean porosity of hydrocarbon bearing rock
S_w = mean water saturation of hydrocarbon bearing rock

$$V_h = \sum_{j=1}^{n} V_{bj} \; \phi_j \; (1 - S_{wj})$$

where V_{bj} = rock volume of porosity ϕ_j and water saturation S_{wj}, and n is the number of segments of different porosity and water saturation required to define the hydrocarbon volume.

Porosity/thickness (ϕh) and *permeability/thickness (Kh)* maps can be created for each zone in a reservoir or for all zones combined. The porosity *(ϕ)* and the gross or net thickness *(h)* for each well is calculated and plotted. These data points are next contoured to generate a ϕh map. Similar procedures are followed to prepare *Kh* (permeability/thickness) map.

Bypassed Oil

In determining waterflood efficiency, we need to calculate the effect of oil migrating from the flooded zones in a reservoir to areas with little or no water. Calculating these effects requires data on the reservoir's volume, pressure and production history.

Recompletion opportunities should be evaluated with an eye toward preventing or recovering trapped oil, and maximizing sweep efficiencies in future operations. These recompletions may involve deepening and/or perforating additional intervals to expose more of the oil zone, or plugging back to reduce excessive water production.

For example, in producing wells that offset, or are adjacent to, injectors, some channeling of injected water may occur, resulting in high water cuts. Injection profile work, followed by the use of plugging material (and, in some cases, selectively plugging back producing wells) may alleviate this problem.

If the differences between reservoir properties in adjacent zones are substantial (i.e., one zone is much more permeable than the other), we may be able to minimize excessive water channeling and reduce the amount of bypassed oil by flooding the zones separately provided they do not vertically communicate with each other. If this approach is economically unattractive, or if the zonal properties variations are minimal, it is common practice to correlate logs (e.g., the gamma-ray and neutron logs in a cased hole) for the injection well with the offset wells, and perforate and/or fracture the wells accordingly.

Geologic Reinterpretation Based On Production History

The reservoir description is the foundation for designing, operating, and evaluating a waterflood or EOR project. It determines (in large part) the selection of a flooding plan and the model for estimating project performance. The development of a reservoir description, which requires extensive computer time and human resources, should start early in the life of a reservoir. It is important to remember that reservoir description is an iterative process. Every description requires modifications as interaction between geologists, geophysicists and engineers working the field matures.

The Elk Basin Madison reservoir provides an example of this iterative process, and illustrates the importance of obtaining extensive reservoir data during field development so that reservoir geology can be defined as soon as possible and incorporated into waterflooding plans.

During the Elk Madison waterflood, a revised reservoir description was used to help interpret the observed production data. In the initial water injection program, water breakthrough was rapid in the interior wells and caused scaling problems, which resulted in production rate declines. In combination with these initial results, a revised description was utilized to alter the water injection program and to drill new producing wells in underdeveloped areas. Figure 3-13 shows the performance history of the Elk Basin Madison resulting from this analysis. This resulted in an increase in ultimate recoverable reserves of 62 million barrels, or 8% OOIP.

The process of developing a reservoir description involves many tasks:

- identifying and mapping the reservoir and non-reservoir rocks
- determining and mapping the rock properties (e.g., porosity, permeability and fluid saturations)
- determining the depositional environment
- determining the number and distribution of reservoir zones

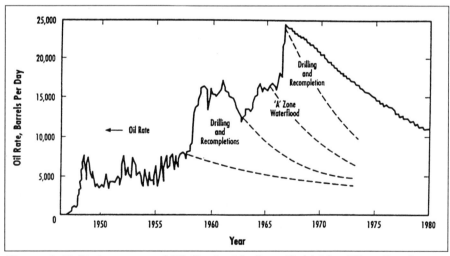

Figure 3-13 Performance of Elk Basin, Madison Field After Waterflooding Plans Were Revised (McCaleb, J.A. "The Role of Geology in Reservoir Exploitation," notes prepared for AAPG Petroleum Reservoir Fundamentals School, 1982)

- establishing the continuity of reservoir and non-reservoir rocks
- testing the reservoir model in light of production performance

The Denver Unit Waterflood in the Wasson San Andres field of West Texas, shown in Figure 3-14, illustrates how geological concepts were used to redesign a waterflood, considering pay discontinuities. This field produces from the San Andres carbonate interval at a depth of about 5,000 ft. The productive interval varies from 300 to 500 ft in thickness. The primary producing mechanism was solution gas drive.

Waterflooding was initiated in this field in 1964, using a peripheral pattern, where water was injected below the oil/water contact (see Figure 3-14). Since the reservoir was considered to be continuous, it was assumed that injecting water at or below the oil/water contact would create an edgewater and/or bottomwater drive. Using this reservoir model, injected water was expected to move laterally and vertically throughout the productive formation.

The peripheral waterflood did not perform as expected. Injectivity was low, since the edge wells selected for water injection often had the poorest-quality reservoir rock. The production wells located far from the injectors failed to respond. The volume of water injected was so low that response was sluggish and limited to the first row of production wells. The combination of inadequate injectivity and poor reservoir continuity over long distances caused the peripheral flood to fail.

The Denver Unit example shows how geology can be reinterpreted in light of production history. Detailed geological studies indicated that the pay interval could be divided into 10 discrete zones. These zones were mapped vertically and laterally over distances of several well locations. The maps identified vertical permeability barriers that would restrict the amount of crossflow between zones.

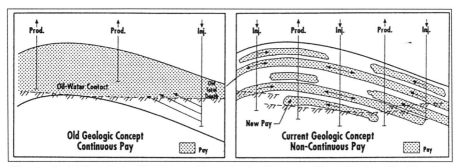

Figure 3-14 Geologic Concepts Used in the Wasson San Andres Field Waterflood (From Ghauri et al., 1980)[17]

These detailed geological studies also showed that a number of pay zones were discontinuous over large distances, and would not be flooded on the 40-acre spacing selected for the waterflood. As a result, a new geological model evolved in which the reservoir was represented as a series of *continuous* and *discontinuous* pay zones. Continuous pay was defined as the portion of the total net pay hydraulically connected between two wells at the field's existing well spacing. Based on this continuity study, the project team began infill drilling on a 20-acre spacing in order to increase the fraction of continuous pay under waterflood.

The design of a primary or enhanced recovery process is generally based on a given geological model of the reservoir. For a displacement process, such as a waterflood or EOR, the volumetric sweep efficiency depends upon contacting as much of the reservoir volume as possible. Flooding plans depend on knowing (or assuming) how pore space is connected between an injection well and a producer, as well as within the reservoir. The reservoir description is generally a geologic model that describes the spatial distribution of fluid and rock properties and saturations for the reservoir and non-reservoir rock within the gross interval and the areal extent of the porous medium.

As we have already mentioned, reservoir description is a dynamic process. The description changes as additional information becomes available. For example, the reservoir description is generally quite simplistic at the time of discovery. It becomes more complex as primary production matures, and infill drilling and/or additional testing (e.g., pressure buildup, drawdown or interference tests) provide data on the average flow characteristics near producing wells. Table 3-8 shows stages in the process of developing a reservoir description.

Our goal should be to define both the reservoir and non-reservoir rock, since non-reservoir rock acts as a barrier to flow. The model is obtained by interpolating the data available at wells—a method that involves both geological and engineering analysis, as well as considerable field experience.

Until recently, a great deal of geological model development was performed by geologists or engineers working independently. Historically, reservoir description has been left to geologists, who usually developed geological maps and cross-sections, and had little interest in (or incentive to) relate their geological descriptions to engineering and production data. Reservoir and production engineers were responsible for developing waterflood plans and predicting performance. Maps and other geological information were used to run simulations and perform various reservoir engineering calculations. During simulation, geological maps were generally modified when they failed to match primary production or waterflood history.

Table 3-8 Flow Path Depicting Development of a Reservoir Description (Thakur, G.C. "Reservoir Management of Mature Fields," IHRDC Video Library for Exploration and Production Specialists, Boston, MA, 1992)

Data	Geologic and Engineering Analysis	Reservoir or Geologic Model
Well logs		Reservoir definition
Drillers logs		Boundaries—areal extent
Cuttings		Oil/water contact, Gas/oil contact
Core material	**Geologic**	Gross thickness
Petrographic and	**and**	Hydrodynamic environment
mineralogic analysis	**Engineering**	Spatial distribution of
Special tests	**Analysis**	properties for reservoir and
Production performance		nonreservoir rock within the
Pressure buildup, draw-		gross thickness
down, and interference		Permeability
tests		Porosity
		Fluid saturations
		Mineralogy
		Continuity
		Heterogeneity

The development of reservoir description has evolved from the need to improve prediction of reservoir simulation models, manage existing waterfloods and design other EOR processes. Developing a reservoir description, like effective reservoir management itself, requires interaction among geologists, geophysicists and engineers. This interaction allows geological assumptions and interpretations to be compared to actual reservoir performance as documented by production history and pressure tests and provides a means of checking the physical properties used in simulators (e.g., porosity, permeability, thickness, zonation, and fluid saturation) to make sure they are consistent with geological interpretations.

Reservoir Heterogeneities

Reservoir heterogeneity is largely dependent upon the depositional environment and subsequent events, along with the nature of sediments involved. The variation in rock properties with evaluation is primarily because of differing depositional environments in time sequence. In sandstone reservoirs, the development of rock properties, e.g., porosity and permeability is dependent on the nature of the sediment, on the depositional environment, and a subsequent compaction and/or cementation. The development of rock properties in carbonates may happen similar to sandstone, along with development as a result of solution, dolomitization, etc. In addition, faulting and fracturing may occur in both types of rocks, leading to more complex reservoir heterogeneities. All these heterogeneities affect the design, implementation, and performance of waterfloods.

Areal and vertical heterogeneities are determined by a combination of geologic, rock and fluid, logging and coring, well testing, and production/injection performance analysis. The presence and direction of fractures critically affect the performance of waterflood, thus their characterization is absolutely necessary early in the life of the reservoir, preferably during primary production.

Figure 3-15 shows a plot of cumulative flow capacity (permeability-thickness) vs. total pore volume (porosity-thickness). It is used to show the contrast in permeabilities, the greater difference indicated by the increased divergence from a 45° line.

The variation of vertical permeability is generally described by the Dykstra-Parsons permeability variation factor. It is based upon the log normal permeability distribution, and statistically it is defined as:

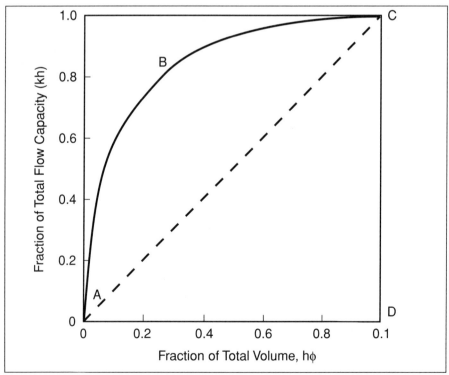

Figure 3-15 Flow Capacity Distribution, Hypothetical Reservoir

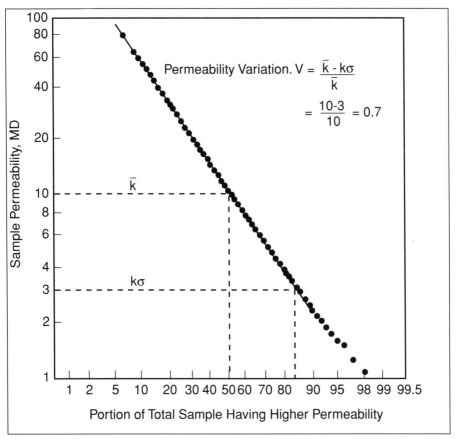

Figure 3-16 Log Normal Permeability Distribution

$$V = \frac{\overline{k} - k_\sigma}{\overline{k}}$$

where

$\overline{k}$ = mean permeability (i.e., permeability at 50% probability)

k_σ = permeability at 84.1% of the cumulative sample

See Figure 3-16 for an example calculation. The permeability variation ranges from 0 (uniform) to 1 (extremely heterogeneous).

The most probable behavior of a heterogeneous system approaches that of a uniform system with a permeability equal to the geometric mean of the samples, i.e.,

$$\bar{k} = (k_1 \cdot k_2 \cdots k_n)^{1/n}$$

An average permeability of several permeabilities in series is

$$\frac{1}{\bar{k}} = \frac{1}{n} \left(\frac{1}{k_1} + \frac{1}{k_2} + \cdots + \frac{1}{k_n} \right)$$

An average permeability of several permeabilities in parallel is

$$\bar{k} = \frac{k_1 + k_2 + \dots + k_n}{n}$$

Intregration of Geoscience and Engineering

Halbouty stated in 1977, "It is the duty and responsibility of industry managers to encourage full coordination of geologists, geophysicists, and petroleum engineers to advance petroleum exploration, development, and production."[1] Despite the emphasis, progress on integration has generally been slow. In recent years, the synergism between the above-mentioned functions have been accelerated as a result of the team approach to reservoir management.

Sessions and Lehman presented the concept of increased interaction between geologists and reservoir engineers through multifunctional teams and cross-training between the disciplines.[2] They stated that production geology and reservoir engineering within the conventional organization function separately, and very seldom does a production geologist get indepth experience in reservoir engineering and vice versa. They advocated cross-exposure and cross-training between disciplines. Integrated reservoir management training for geoscientists and engineers offered by many major oil and gas companies is designed to address these needs.[3]

Sessions and Lehman presented Exxon's three case histories where the geology-reservoir engineering relationship was promoted through both a team approach and an individual approach. The results of the three cases (project-based approach, team-based approach, and multiskilled individual approach) were very positive.

Synergy and team concepts are the essential elements for integration of geoscience and engineering.[4] It involves people, technology, tools, and data. Success for integration depends on

- Overall understanding of the reservoir management process, technology, and tools through integrated training and integrated job assignments
- Openness, flexibility, communication, and coordination
- Working as a team
- Persistence

Reservoir engineers and geologists are beginning to benefit from seismic and cross-hole seismology data. Also, it is essential that geological and engineering ideas and reasoning be incorporated into all seismic results if the full economic value of the seismic data is to be realized.

Perfectly conscientious and capable seismologists may overlook a possible extension in a proven area because of their unfamiliarity with the detailed geology and engineering data obtained through development.

For this reason, geological and engineering data should be reviewed and coordinated with the geophysicists to determine whether or not an extension is possible for the drilling of an exploratory well. Most of the difficulties encountered in incorporating geological and engineering knowledge into seismic results and vice versa may be averted by an exchange of these ideas between the three disciplines.

Robertson of Arco points out that the geologic detail needed to properly develop most hydrocarbon reservoirs substantially exceeds the detail required to find them.[5] This perception has accelerated the application of 3D-seismic analysis to reservoir management. A 3D-seismic analysis can lead to identification of reserves that may not be produced optimally (or perhaps not produced at all) by the existing reservoir management plan. In addition, it can save costs by minimizing dry holes and poor producers.

The initial interpretation of a 3D-seismic survey affects the original development plan. With the development of the field, additional information is collected and is used to revise and refine the original interpretation. Note that the usefulness of a 3D-seismic survey lasts for the life of the reservoir.

The geophysicists' interpretation of the 3D-seismic data may be combined with the other relevant information regarding the reservoir (i.e., trap, fault, fracture pattern, shapes of the deposits). The 3-seismic data guide interwell interpolations of reservoir properties. The reservoir engineer can use the seismic volume to understand lateral changes.

The 3D-seismic analysis can be used to look at the flow of fluids in a reservoir. Such flow surveillance is possible by acquiring baseline 3D-data before and after the fluid flow and pressure/temperature changes. Although flow surveillance with multiple 3D-seismic surveys is at an early stage of application, it has been successfully applied in thermal recovery projects.

Cross-well seismic tomography is developing into an important tool for reservoir management, and within the last few years there have been notable advances in the understanding of the imaging capability of crosswell tomograms. The fundamental requirements for the technology have been demonstrated. High-frequency seismic waves capable of traveling long interwell distances can be generated without damaging the borehole, and tomographic inversion techniques can give reliable images as long as the problems associated with nonuniform and incomplete sampling are handled correctly.

Cross-well seismology is becoming an important tool in reservoir management. Current applications focus on the monitoring of enhanced oil recovery processes, but perhaps most important is the potential of the method to improve our geological knowledge of the reservoir.

The role of geology in reservoir simulation studies was well described by Harris in 1975[6]. He described the geological activities required for constructing realistic mathematical reservoir models. These models are used increasingly to evaluate both new and mature fields and to determine the most efficient management scheme. Part of the information contained in the model is provided by the geologist, based on studies of the physical framework of the reservoir. However, for the studies to be useful the geologist must develop quantitative data. It is important that the geologist and the engineer understand each other's data.[6]

As described by Harris, both engineering and geological judgment must guide the development and use of the simulation model. The geologist usually concentrates on the rock attributes in four stages:

1. Rock studies establish lithology and determine depositional environment, and reservoir rock is distinguished from non-reservoir rock
2. Framework studies establish the structural style and determine the three-dimensional continuity character and gross-thickness trends of the reservoir rock

3. Reservoir-quality studies determine the framework variability of the reservoir rock in terms of porosity, permeability, and capillary properties (the aquifer surrounding the field is similarly studied)

4. Integration studies develop the hydrocarbon pore volume and fluid transmissibility patterns in three dimensions

Throughout his work, the geologist requires input and feedback from the engineer. Examples of this "interplay of effort" are indicated in Figure A-11.[6] Core-analysis measurements of samples selected by the geologist provide data for the preliminary identification of reservoir rock types. Well test studies aid in recognizing flow barriers, fractures, and variations in permeability. Various simulation studies can be used to test the physical model against pressure-production performance; adjustments are made to the model until a match is achieved.

Many companies have initiated the development of a three-dimensional geological modeling program to automate the generation of geologic maps and cross-sections from exploration data. A good example of putting geology into reservoir simulations is described by Johnson and Jones.[7] The models are directly interfaced to the reservoir simulator; thus, the reservoir engineer utilizes the complex reservoir description provided by the geologist for field development planning. The reservoir engineer routinely and readily updates the model with new data or interpretations and quickly provides consistent maps and cross-sections.

According to Johnson and Jones, the geologist can input structural and stratigraphic concepts as a series of computer grids honoring the geologic tops. Interpolations of logged porosity and other data from wells are controlled by this stratigraphic framework and fill a 3-D matrix of cells. Geological features critical to reservoir performance can be added to complete the geologist's picture of the reservoir.

Recently, Frank, Van Reet, and Jackson presented an excellent example of synergistic combination of geostatistics, 3-D seismic data, and well log data which contributed to the success in pinpointing infill drilling targets in Kingdom Abo field in Terry County, Texas.[8] In this project, geostatistics is a powerful tool for reservoir characterization, since it utilizes all well data in a manner that adheres to a model based on statistical and user-defined spatial correlations. When coupled with 3-D seismic, the end product is an interpretation of the reservoir that can be used to pinpoint additional development drilling locations.

Integrating Exploration and Development Technology

New developments in computer hardware, technology, and software are enhancing integration of multidisciplinary skills and activities. The mainframe supercomputers, more powerful personal computers, and workstations have revolutionized interdiscipline technical activities and industry business practices, making them more responsive and effective.

Oil and Gas Journal published a special report on "Integrating Exploration and Development Technology" using state-of-the-art computing and communications.[9-12] The *OGJ* special report states that integration is changing the way oil companies work. However, integration also creates challenges, from managing computer systems to designing organizations, to making best use of interdiscipline teams.

Neff and Thrasher captured the significant impacts that the late 1970s and 1980s new technologies made on the petroleum industry.[9] Major computer technologies include the supercomputers, interactive workstations, networking, rapid access mass-storage devices, and 3-D visualization hardware. These resources are enhancing the integration of the activities of the multidisciplinary groups by utilizing their own professional technologies, tools, and data.

Advancements in 3-D seismic acquisition and processing are credited to the massive number-crunching supercomputers such as Cray computers. 3-D seismic data along with computer-processed logs and core analyses characterize or describe more realistically and accurately the reservoir providing the 3-D computer maps. The reservoir engineers use these maps along with rock and fluid properties and production/injection data to simulate reservoir performance and to design depletion and development strategies for new and old fields. The supercomputers made reservoir simulators work faster and more accurately. The integration process from reservoir characterization to reservoir simulation, which requires interdisciplinary teamwork has been made practical and efficient by utilization of computers.

Interactive workstations interface several machines together locally in a physical cluster or using networks and software to link central processing units (CPUs) from various sites into a virtual cluster. The machines include high-end PCs, Suns, DECs, IBMs, MicroVAXes, Hewlett-Packard (HPs), and Silicon Graphics hardware. Contrary to the workings of the supercomputers and mainframe computers, the interactive workstations allow data migration, analysis, and interpretation on truly interactive domain rather than batch mode. The workstations are also capable of utilizing many geoscience and engineering software interactively. The demands for workstations of various kinds are ever increasing in the industry because they are becoming the workhorse of the integrated geoscience and engineering teams.

The computer networks that link the IBM mainframe computers, Cray supercomputers, UNIX workstations, and PC token ring networks together provide the mechanism for effective communication and coordination from various geographical office locations. Major oil companies have worldwide computer links between all divisions and regional offices. The office-to-office communication has been made very quick (almost instantaneous), productive, and cost-effective by computer networking. The IBM mainframe-based PROFS/Office Vision electronic mail facilities, videoconference centers in various geographical locations, and workstations' images of maps, graphs, and reports via network communications are excellent examples of networking. The networks have made the tasks of the integrated teams easier, faster, and immensely productive.

While networks provide an efficient means to move digital data, retrieval and storing of data pose a major challenge in the petroleum industry today. The problems are:

- Incompatibility of the software and data sets from the different disciplines
- Databases usually do not communicate with each other

Many oil companies are staging an integrated approach to solving these problems.[12] In late 1990, several major domestic and international/overseas oil companies formed Petrotechnical Open Software Corporation (POSC) to establish industry standards and a common set of rules for applications and data systems within the industry. POSC's technical objective is to provide a common set of specifications for computing systems, which will allow data to flow smoothly between products from different organizations and will allow users to move smoothly from one application to another. POSC members are counting on POSC and its major software vendors to provide a long-term solution to database-related issues.

The latest major breakthrough in computer technology is 3-D computer visualization via a video monitor of a reservoir at a micro-or macro-scale. The awesome power of visualization lies in its ability to synthesize diverse data types, e.g., geology, land, geophysics, petrophysics, drilling, and reservoir engineering, and attributes for better understanding and capturing by human senses. Figure 3-17 is an example of visualization of a Gulf Coast Mexico salt dome, blending many

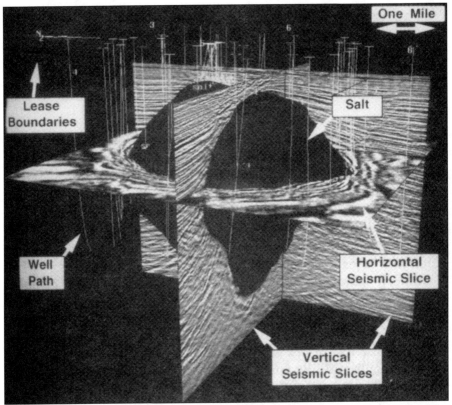

Figure 3-17 Integrating Exploration and Development Technology (Courtesy OGJ, May 1993[9], from Wyatt et al., "Ergonomics in 3-D depth migration," 62nd SEG Int. Mtg. And Exp., October 1992)

types of information. Figure 3-18 shows computer visualization of electron microscope pictures of rock samples alongside classic rock displays. 3-D visualization technique will enhance our understanding of the reservoir, providing better reservoir description and simulation of reservoir performance. It may very well be the most powerful and persuasive communication tool of the integrated teams for decades to come.

Now, a time and cost-effective way to integrate exploration and production activities using existing hardware and software is available.[10] A fully open-data exchange system, which was jointly created by Finder Graphics Systems Inc., GeoQuest, and Schlumberger, is being distributed as the Geoshare standards. Members of the Geoshare user's group, which consists of many geoscientific software developers and oil and gas operators, will soon be able to transfer data and interpretations among their various data bases in support of E&P techniques.

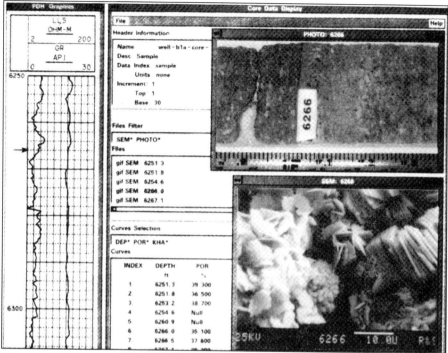

Figure 3-18 Integrating Exploration and Development Technology (Courtesy OGJ, May 1993[9])

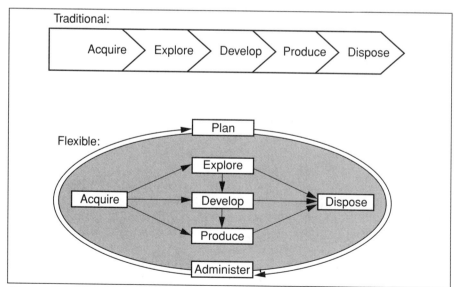

Figure 3-19 Two Types of Organization (Copyright 1993, Gemini Consulting, courtesy OGJ, May 1993[11])

Guthery, Landgren, and Breedlove concluded that the published Geoshare standard provides means for exchanging data and results between any petroleum applications, regardless of their formats, configurations or hardware platforms.[10] It is a completely open and expandable standard whose future lies with the Geoshare user's group.

Traditionally, finding and producing hydrocarbons were considered the essence of success in the upstream end of the petroleum industry. Now, companies are viewing their options as far more flexible, and a diversified portfolio of skills within an integrated and flexible business framework is emerging (Figure 3-19).[11] Patterson and Altieri discussed two dominant approaches to convert the "stovepipe" organization of separate functional units into an integrated organization.

In the "top-down" first approach, changes in strategy and management style made at the executive level are expected to filter down and throughout the organization over time. The use of asset management teams is now standard practice in many companies. Even though this kind of teamwork and flexibility is a step in the right direction, it does not really address broader organization and information technology issues. In the emerging second approach, which works from the bottom up, changes in information technology alone (both hardware and software) are intended to eliminate problems of knowledge transfer and communication.

While significant business improvements may be produced from the best of both approaches, neither provides the blueprints necessary for achieving the linkage between a company's strategic direction and its organization, operations, and information systems. Patterson and Altieri found that by modeling the "what" (the work of the business) and "why" (the purpose of the work), it is possible to build a stable blueprint that can be used to redesign and align the entire business. Once the fundamental purpose of business is defined, the modeling of the work that supports the purpose is approached first at the middle levels of the business. Then from the middle level, the analysis is driven in both directions—upward, to get higher levels of work, and downward, to lower levels. This process will produce a detailed blueprint of the essential work required to achieve the business purpose and of the information required to accomplish the work.

The work needs to be carried out by cross-functional teams with a common objective and smooth line of communication between the different functional groups of the organization. The team maps out the actual work flows, comparing them to the work defined in the completed model, which serves as the zero base. Since the team members analyze the business from the standpoint of actual work requirements (not functional assignments), they are able to see that individuals from different functional units are contributing to the same essential work.

Using the model as the basic framework, the old "stove pipe" organization can be converted into an integrated organization in which crossfunctional teams focus entirely on work that supports the objectives of the business. Everyone involved gains, not just an individual functional area. The team members share a common business objective, which effectively eliminates most political disputes, functional rivalries, and fear of change. The final result is a new, seamless organization that is flexible and adaptable to change. It can quickly move the focus of its work anywhere within the business life cycle to maximize value creation in an ever changing marketplace.

Development Of Reservoir Description

Table 3-9 describes activity, input parameters, objectives and primary responsibility for the development of reservoir description. As mentioned earlier is this chapter, the process of developing a reservoir description involves many tasks:

- identifying and mapping the reservoir and non reservoir rocks
- determining and mapping the rock properties (e.g., porosity, permeability and fluid saturations)
- determining the number and distribution of reservoir zones
- establishing the continuity of reservoir and non-reservoir rocks
- testing the reservoir model in light of production performance

It is understood that, in spite of careful and methodical work, reservoir description is not very precise because: 1) all facts are not available, 2) ideas or perceptions change with new factors or experience, 3) interpretations are probably more simplified than is actually the case (i.e., the earth is complex), 4) interpretations usually change with time. Levorsen pointed out very clearly that "one thing to bear in mind with regard to subsurface maps is that they are never finished.[13] They may be thought of as progress maps or contemporary maps, only as complete as the data that are available when they are made."

Determining the correct geologic model is a complicated effort that must consider many factors: 1) deposition of sediments, 2) general and specific features, 3) structural deformation, 4) structural alterations, 5) proper sequence of geologic events. A detailed description involving the concept of reservoir modeling is given in Appendix B.

Table 3-9 Development of a Reservoir Description (Copyright 1975, SPE, from JPT, May 1975 and SPE Textbook[6] Series, Vol. 3, 1986[15])

Activity	Input Parameters	Objectives	Primary Responsibility
Rock studies	Cores	Core descriptions—including mineralogy	Geologist
	Cuttings	Lithology (facies)	Geologist
	Core analysis	Depositional environment	Geologist
	Drillers logs	Correlations of core data for each	Engineer/Geologist
	Well logs	productive facies	
		Porosity — k_h	
		k_h — k_v	
		S_w — k_h	
Framework studies	Core analyses	Log porosity—core porosity	Geologist/Geophysicist
	Core description graphs	Structure map	Geologist
	Well logs	Delineation of aquifer/gas cap	
	Depositional model	Principal reservoir zones	
		Areal and vertical extent of principal	
		reservoir/nonreservoir zones	
	Correlations of rock	Continuity of reservoir and non	Geologist/Engineer
	properties	reservoir zones	
	Pulse tests	Cross sections	
	Seismic data	Fence or panel diagrams	
Reservoir quality studies	Framework maps	Area distribution of permeability,	Engineer/Geologist
	Core analyses	porosity, net sand and water	Geologist/Engineer
	Well logs	saturation for each reservoir zone	
	Correlations of rock	Contour maps of permeability, porosity,	
	properties	water saturation, and net sand	
	Well tests—pulse,		
	buildup, and falloff		
Integration studies	Reservoir description from	Reservoir maps	Engineer/Geologist
	rock and reservoir quality	PV	
	studies	Hydrocarbon PV	
		Transmissibility (k_h)	
	Mathematical model	History match of pressure/production	Engineer
	simulating production/	Response by adjusting rock parameters	Geologist
	pressure primary	Consistent with geological model	
	production waterflood		

References

1. Halbouty, M. T. "Synergy is Essential to Maximum Recovery," *JPT* (July 1977): 750-754.
2. Sessions, K. P. and D. H. Lehman. "Nurturing the Geology—Reservoir Engineering Team: Vital for Efficient Oil and Gas Recovery." SPE Paper 19780 presented at the Annual Technical Conference and Exhibition," San Antonio, TX, October 8-11, 1989.
3. Satter, A. "Reservoir Management Training—An Integrated Approach." SPE Paper 20752, Reservoir Management Panel Discussion, SPE 65th Annual Technical Conference & Exhibition, New Orleans, LA, September 75): 625-632.
4. Satter, A. and G. C. Thakur. "Integrated Petroleum Reservoir Management: A Team Approach," *PennWell Books*, Tulsa, Oklahoma, 1994.
5. Robertson, J. D. "Reservoir Management Using 3-D Seismic Data," *Geophysics: The Leading Edge of Exploration* (February 1989): 25-31.
6. Harris, D. G. "The Role of Geology in Reservoir Simulation Studies," *JPT* (May 1975): 625-632.
7. Johnson, C. R. and T. A. Jones. "Putting Geology Into Reservoir Simulations: A Three-Dimensional Modeling Approach." SPE Paper 18321, presented at the Annual Technical Conference and Exhibition, Houston, TX, October 2-5, 1988: 585-594.
8. Frank, Jr., J. R, E. Van Reet and W. D. Jackson. "Combining Data Helps Pinpoint Infill Drilling Targets in Texas Field," *Oil & Gas J.* (May 31,1993): 48-53.
9. Neff, D. B. and T. S. Thrasher. "Technology Enhances Integrated Teams Use of Physical Resources," *Oil & Gas J.* (May 31, 1993): 2-35.
10. Guthery, S., K. Landgren and J. Breedlove. "Data Exchange Standard Smooths E&P Integration," *Oil & Gas J.* (May 31, 1993): 36-42.
11. Patterson, S. and J. Altieri. "Business Modeling Provides Focus for Upstream Integration," *Oil & Gas J.* (May 31, 1993): 43-47.
12. Johnson, J. P. "POSC Seeking Industry Software Standards, Smooth Data Exchange," *Oil & Gas J.* (October 26, 1992): 64-68.
13. Levorsen, A. L. "Geology of Petroleum," 2nd ed., W. H. Freeman & Co., September 1967.
14. Thakur, G. C. "Reservoir Management of Mature Fields," IHRDC Video Library for Exploration and Production Specialists, Boston, MA (1992).
15. Willhite, G. P. "Waterflooding," SPE Textbook Series, Vol. 3 (1986).
16. McCaleb, J. A. "The Role of Geology in Reservoir Exploitation," notes prepared for AAPG Petroleum Reservoir Fundamentals School (1982).
17. Ghauri, W. K. "Production Technology Experience in a Large Carbonate Waterflood, Denver Unit, Wasson San Andres Field," *JPT* (September 1980), 1493-1502.

4

Waterflood Data

Throughout the life of a reservoir, from exploration to abandonment, a great amount of multidisciplinary data are collected (Fig. A-1).[1] Figure 4-1 shows a list of data collected before and during production. An integrated approach involving all functions is necessary to develop an efficient data management program, which plays a key role in reservoir management[1]. It requires planning, justification, prioritizing and timing to lay down the foundation of reservoir management.

Appendix C presents the basic geoscience and engineering data required for reservoir management. The appendix also presents a general discussion on data acquisition, analysis, validation, storing and retrieval, application, and examples.[1] This chapter presents specific laboratory and field data required for waterflood asset management.

Laboratory Data

Laboratory data needed for waterflood are:

- Fluid properties
- Rock properties
- Flow properties
- Residual oil saturation
- Water quality

Fluid Properties

The fluids contained within a reservoir are usually water, gas, and liquid hydrocarbons. Their properties are highly dependent upon the reservoir pressure and temperature in addition to their composition. The composition and properties of the fluids are determined by laboratory

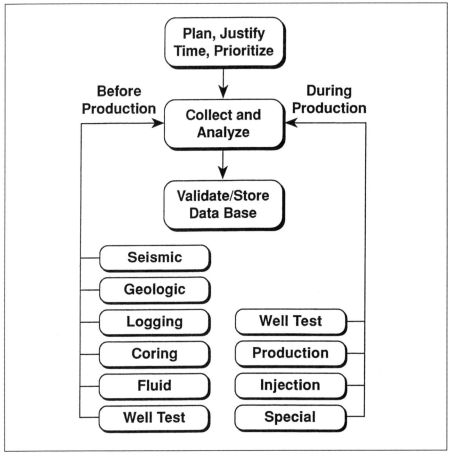

Figure 4-1 Data Acquisition and Analysis (Satter and Thakur, *Integrated Petroleum Reservoir Management: A Team Approach*, PennWell, 1994)

tests, depending upon the type of fluids present, and the sampling methods used, i.e., equilibrium flash or differential liberation. In the flash liberation test, the reservoir oil, either a subsurface sample or a recombination of surface samples from separators and stock tanks, is charged into a pressure vessel at reservoir temperature. The pressure is then lowered in small increments, and the volumes of the liquid and the liberated gas below bubble point are recorded. Differential liberation is conducted so that the solution gas is continuously removed from the system.

The results of differential vaporization tests conducted by Core Laboratories on a 40.5° API crude at 220°F with a solution gas-oil ratio of 854 scf/bbl and 2620 psig bubble point are shown in Tables 4-1 to 4-3, and Figures 4-2 to 4-8. [2,3]

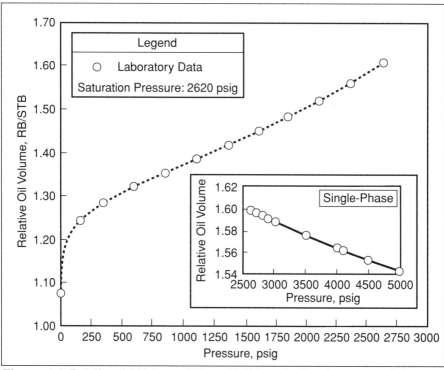

Figure 4-2 Relative Oil Volume Differential Vaporization (courtesy of Core Laboratories Inc.)

Rock Properties

The basic rock properties are concerned with the properties of the rock material alone, and the properties related to the interaction of the rock and the fluids contained in the pores. Rock properties include porosity, pore size distribution, permeability, compressibility and surface area, which are determined by laboratory tests. Rock-fluid interaction include:

- Rock wettability
- Capillary pressure
- Relative permeability

Table 4-1 Composition of Reservoir Fluid Sample (by Flash/Extended Chromatography, courtesy of Core Laboratories Inc.)

Component Name	Mol %	Wt %	Liquid Density (gm/cc)	MW
Hydrogen Sulfide	0.00	0.00	0.8006	34.08
Carbon Dioxide	0.91	0.43	0.8172	44.01
Nitrogen	0.16	0.05	0.8086	28.013
Methane	36.47	6.25	0.2997	16.043
Ethane	9.67	3.10	0.3558	30.07
Propane	6.95	3.27	0.5065	44.097
iso-Butane	1.44	0.89	0.5623	58.123
n-Butane	3.93	2.44	0.5834	58.123
iso-Pentane	1.44	1.11	0.6241	72.15
n-Pentane	1.41	1.09	0.6305	72.15
Hexanes	4.33	3.88	0.6850	84
Heptanes	2.88	2.95	0.7220	96
Octanes	3.15	3.60	0.7450	107
Nonanes	2.16	2.79	0.7640	121
Decanes	2.12	3.03	0.7780	134
Undecanes	1.85	2.90	0.7890	147
Dodecanes	1.61	2.77	0.8000	161
Tridecanes	1.40	2.62	0.8110	175
Tetradecanes	1.23	2.50	0.8220	190
Pentadecanes	1.08	2.38	0.8320	206
Hexadecanes	0.95	2.25	0.8390	222
Heptadecanes	0.85	2.15	0.8470	237
Octadecanes	0.76	2.04	0.8520	251
Nonadecanes	0.70	1.97	0.8570	263
Eicosanes plus	12.55	43.54	0.9004	325
Totals	100.00	100.00		

Total Sample Properties

Molecular Weight 93.66

Equivalent Liquid Density, gm/scc 0.7015

Plus Fractions	Mol %	Wt %	Density	MW
Heptanes plus	33.29	77.49	0.8515	218
Undecanes plus	22.98	65.12	0.8736	265
Pentadecanes plus	16.89	54.33	0.8887	301
Eicosanes plus	12.55	43.54	0.9004	325

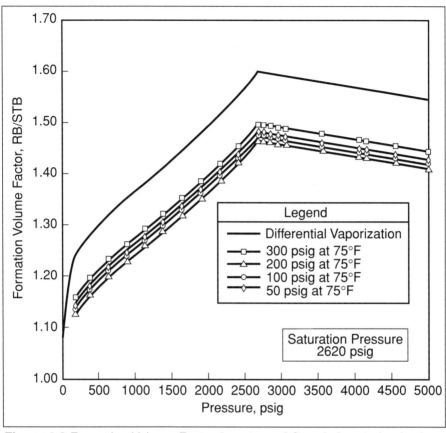

Figure 4-3 Formation Volume Factor (courtesy of Core Laboratories Inc.)

Table 4-2 Volumetric Data (courtesy of Core Laboratories Inc.)

Saturation Pressure (Psat)........................2620 psig
Density at Psat...0.6562 gm/cc
Thermal Exp @ 5000 psig ....1.08795 V at 220°F/V at 60°F

Average Single-Phase Compressibilities		
Pressure Range **psig**		**Single-Phase Compressibility** **v/v/psi**
5000 to 4500		13.14 E-6
4500 to 4000		14.08 E-6
4000 to 3500		15.26 E-6
3500 to 2800		17.31 E-6
2800 to 2620		19.92 E-6

Table 4-3 Separator Flash Analysis (courtesy of Core Laboratories Inc.)

Flash Conditions psig / °F	Gas/Oil Ratio (scf/bbl) (A)	Gas/Oil Ratio (scf/STbbl) (B)	Stock Tank Oil Gravity at 60°F (°API)	Formation Volume Factor Bofb (C)	Separator Volume Factor (D)	Specific Gravity of Flashed Gas (Air=1.000)	Oil Phase Density (gm/cc)
300	478	549			1.148	0.704*	0.7771
0	244	246	40.1	1.495	1.007	1.286	0.8177
		795 Rsfb=					
200	542	602			1.112	0.732*	0.7856
0	177	178	40.4	1.483	1.007	1.329	0.8170
		780 Rsfb=					
100	637	676			1.062	0.786*	0.7985
0	91	92	40.7	1.474	1.007	1.363	0.8156
		768 Rsfb=					
50	715	737			1.031	0.840*	0.8088
0	41	41	40.5	1.481	1.007	1.338	0.8166
		778 Rsfb=					

* Collected and analyzed in the laboratory by gas chromatography.
(A) Cubic Feet of gas at 14.65 psia and 60°F per Barrel of oil at indicated pressure and temperature.
(B) Cubic Feet of gas at 14.65 psia and 60°F per Barrel of Stock Tank Oil at 60°F.
(C) Barrels of saturated oil at 2620 psig and 220°F per Barrel of Stock Tank Oil at 60°F.
(D) Barrels of oil at indicated pressure and temperature per Barrel of Stock Tank Oil at 60°F.

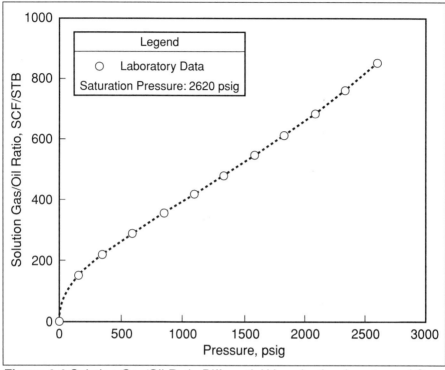

Figure 4-4 Solution Gas/Oil Ratio Differential Vaporization (courtesy of Core Laboratories Inc.)

Rock Wettability

An oil-water-solid system can be either water or oil-wet depending upon the nature of the rocks and the types of organic materials present. In a strongly water-wet system, water spreads on or adheres to the rock; in the oil-wet system, oil adheres to the rock. The wettability influences capillary pressure and relative permeability characteristics which control oil recovery due to waterflooding.

Wettability can be experimentally determined by contact angle tests on pure quartz or carbonate crystal. The test consists in measuring through the water the angle at the oil-water-solid interface. Test data yield changing contact angles with time, and sometimes require two weeks or more to stabilize.[2] Contact angles in the vicinity of zero and 180° are considered strongly water-wet and strongly oil-wet, respectively. Contact angles near 90° are "intermediate wettability".[4]

Another technique due to Amott wettability index is based upon the premise that a strongly wetting fluid will spontaneously imbibe until

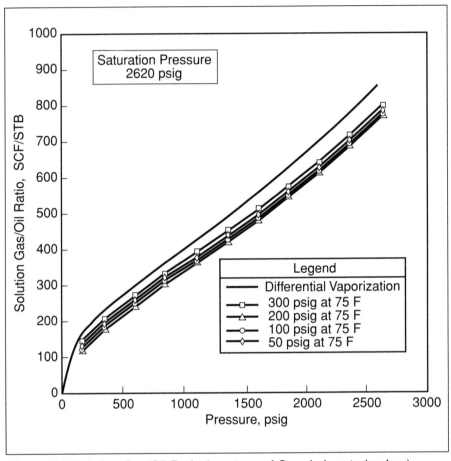

Figure 4-5 Solution Gas/Oil Ratio (courtesy of Core Laboratories Inc.)

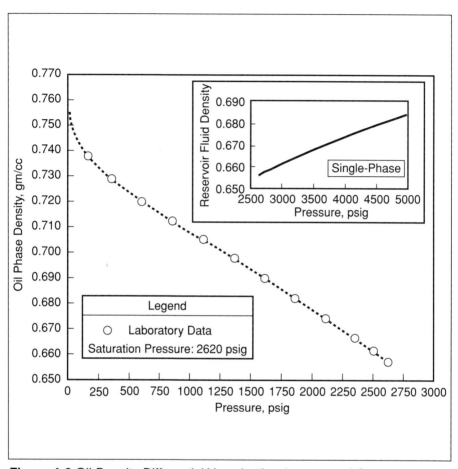

Figure 4-6 Oil Density Differential Vaporization (courtesy of Core Laboratories Inc.)

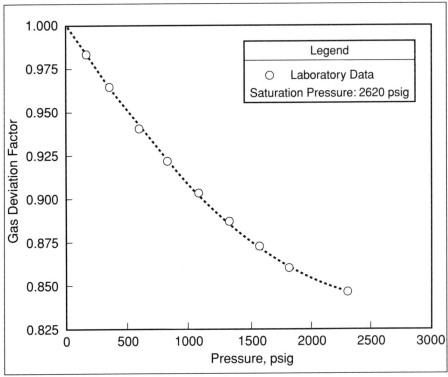

Figure 4-7 Deviation Factor, Z Differential Vaporization (courtesy of Core Laboratories Inc.)

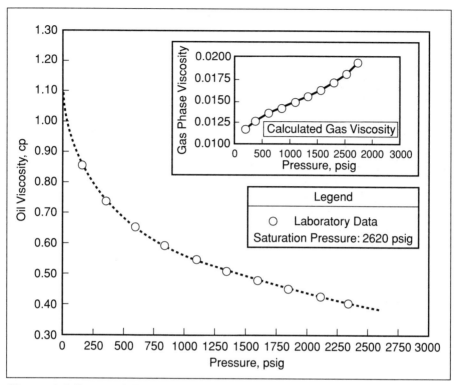

Figure 4-8 Reservoir Fluid Viscosity (courtesy of Core Laboratories Inc.)

the residual saturation of the non-wetting fluid is obtained.[5] An index of 1.0 indicates a strongly wetting fluid, an index of 0.0 a strongly non-wetting fluid.

Capillary Pressure

The capillary pressure is defined as the pressure in the non-wetting phase minus the pressure in the wetting phase, generally a positive value.[4] Then the water-wet capillary pressure is the pressure in the oil phase minus the pressure in the water phase, and the oil-wet capillary pressure is the pressure in the water phase minus the pressure in the oil phase. Figures 4-9 and 4-10 present the capillary pressure characteristics of a strongly water-wet and oil-wet system, respectively.[6] Note that the capillary pressures are influenced by the drainage and the imbibition processes. Figure 4-11 presents results of air-brine capillary pressure test on a sandstone core with 10% porosity and 560 md air permeability.

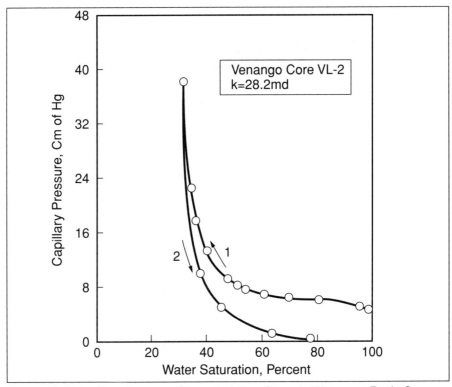

Figure 4-9 Capillary Pressure Characteristics, Strongly Water-wet Rock. Curve 1-Drainage. Curve 2-Imbibition (Killias, C.R, et al., *Production Monthly*, Feb. 1953)

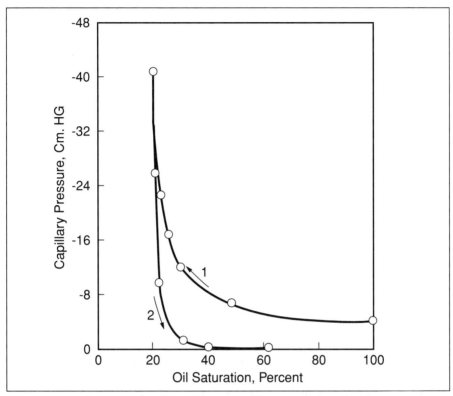

Figure 4-10 Oil-water Capillary Pressure Characteristics, Tensleep Sandstone, Oil-wet Rock. Curve 1-Drainage. Curve 2-Imbibition (Killias, C.R, et al., *Production Monthly*, Feb. 1953)

Relative Permeability

Relative permeability is defined as the ratio of the effective permeability of a fluid at a given saturation to some reference permeability, such as the absolute permeability of the rock or the effective permeability of the fluid at irreducible water saturation. Absolute permeability is a measure of the ability of the rock to transmit any fluid when the rock is 100% saturated with that fluid. The effective permeability is a measure of the conductance of the porous media for one fluid phase when the media is saturated with more than one fluid.

Typical water-oil relative permeabilities are presented for water-wet and also for oil-wet formations in Figures 2-2 and 2-3, respectively.[4] Figures 4-12 and 4-13 present unsteady state gas-oil, and water-oil relative permeability characteristics, respectively, of a sandstone core with 10% porosity and 560 md air permeability.

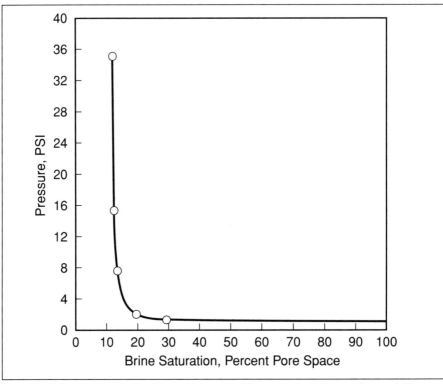

Figure 4-11 Air-brine Capillary Pressure Characteristics (courtesy of Core Laboratories Inc.)

The following rules of thumb indicate the differences in the flow characteristics due to rock wettability:[4]

	Water-Wet	**Oil-Wet**
Connate water saturation	Usually greater than 20 to 25% pore Volume	Generally less than 15% pore volume, frequently less than 10%
Saturation at which oil and water relative permeabilities are equal	Greater than 50% water saturation	Less than 50% water saturation
Relative permeability to water at maximum water saturation; i.e., floodout	Generally less than 30%	Greater than 50% and approaching 100%

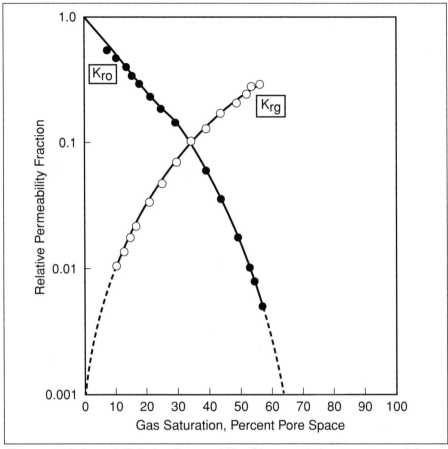

Figure 4-12 Gas-oil Relative Permeability Characteristics (courtesy of Core Laboratories Inc.)

It has generally been found that rocks with intermediate wettability have some of the above characteristics of both water-wet and oil-wet formations. Most formations are of intermediate wettability, i.e., no strong preference for either water or well.

Residual Oil Saturation

Residual oil saturation (or water saturation at no oil flow), which is a measure of the waterflood recovery efficiency can be estimated directly from the relative permeability curves (see Figures 2-2 and 2-3).

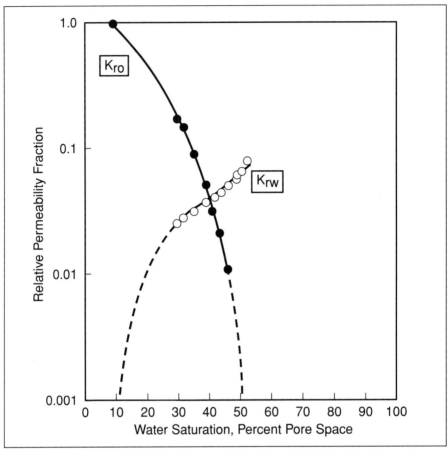

Figure 4-13 Water-oil Relative Permeability Characteristics (courtesy of Core Laboratories Inc.)

Laboratory flood pot tests can be also carried out to determine residual oil saturation to waterflood.[4] Water is injected into the central hole of a mounted cylindrical whole core until a negligible oil rate or an excessive water-oil ratio is obtained. The residual oil saturation at floodout as well as the volume of water needed for the recoverable oil can be determined from the test results.

Figure 4-14 presents results of waterflood susceptibility test on sandstone core with 10% porosity, 580 md air permeability, and 7.8% initial water saturation. At terminal condition of floodout (99.9% water-cut), oil recovery was 47.6% pore volume.

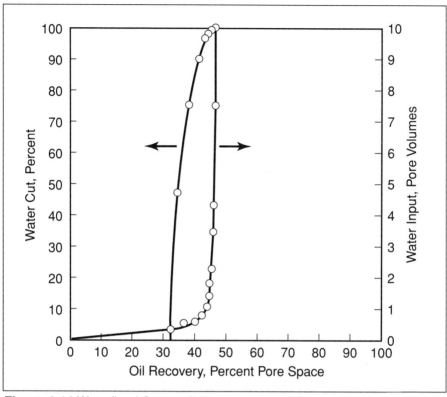

Figure 4-14 Waterflood Susceptibility (courtesy of Core Laboratories Inc.)

Water Quality

Water quality needs to be checked out by:

- Concentration analyses
- Compatibility tests
- Formation sensitivity tests

Laboratory source-water analyses provides the concentration of the dissolved solids such as sodium, calcium, magnesium, barium, iron, chlorides, bicarbonates, carbonate, and sulfates.[7] Specific gravity, resistivity, and pH should be also measured. Suspended solids, oil and grease concentrations, and dissolved gases are generally tested on site. The results of the analyses provide scaling tendencies and corrosivity of the water which should be dealt with.

The reservoir formation water is considered to be compatible with the source water if no precipitate is formed when they are mixed.[7] Treatment should be considered for incompatibility which can create severe plugging and scale deposition problems. However, some waterfloods can be successful, even though the source water is not compatible with the formation water.[8]

Some reservoir rocks, particularly sandstones, suffer permeability reduction when contacted by injected water.[7] Damage which is usually associated with low-salinity injected water is caused by water reacting with the indigenous clays, causing them to swell. Also, as water flows through the rock, dispersing and rearranging fine formation particles cause blocking or plugging of the pore system. Figure 4-15 presents results of water sensitivity test on a sandstone core with 20% porosity and 39.5 md air permeability.

Solutions for mitigating or eliminating the formation damage involve chemical treatments. Most reservoirs may not be water sensitive and may not suffer from permeability damage even if fresh water is injected.

Field Data

In addition to the basic geoscience and engineering data including production and injection (see Appendix B), special tests are needed for efficient waterflood operations:

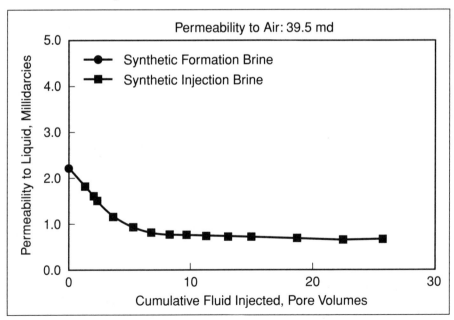

Figure 4-15 Permeability vs. Throughput (courtesy of Core Laboratories Inc.)

- Water injectivity
- Profile surveys
- Transient well pressures
- Tracer surveys

Water Injectivity

Water injectivity is defined as the injection rate per unit pressure difference between the bottom hole injection pressure and the reservoir pressure. Injectivity test is carried out by injecting water into a well at varying rates while measuring the bottom hole pressures. When injection rate is plotted against the bottom hole pressure, a break in the curve results at the parting or formation fracture pressure (Figure 4-16).[7] The parting pressure gradient, normally 1 psi/ft of the well depth, may vary from 0.75 to 1.2 psi/ft. Above the parting pressure, water can be injected at higher rates per increment of pressure.

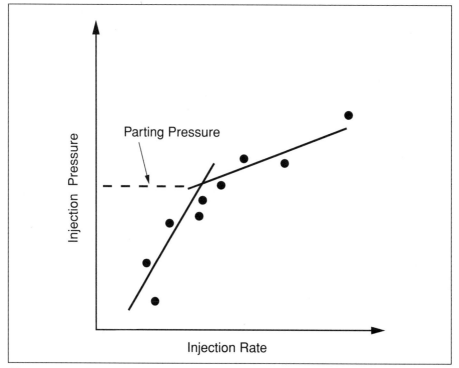

Figure 4-16 Parting-pressure Diagram (Rose, S. C., J. F. Buckwalter, and R. J. Woodhall. "The Design Engineering Aspects of Waterflooding," SPE Monograph 11, Richardson, TX (1989)

Note that the parting pressure is also a function of the reservoir pressure, and at low reservoir pressures the formation can easily be fractured. A rule-of-thumb to follow is that for every 100 psi change in formation pressure, parting pressure changes by 60-70 psi.

It is very important to determine the required injection rate which controls the life of the waterflood project. Normally, injection pressures are set at slightly below the parting pressure. In some special cases, where formation permeabilities are very low and injection water is very clean, injection may be carried on above the parting pressure. If this procedure is followed, it is advisable to run injection profiles regularly to make sure the water injection is contained in zones of interest.

Profile Surveys

Profile surveys are designed to determine the distribution of fluids from the wellbore to the reservoir (injection profile) and from the reservoir to the wellbore (production profile). Spinner flowmeter, radioactive tracer and temperature surveys are commonly run to determine the distribution of injected fluids during injection. Figure 4-17 illustrates a typical injection well profile.

Periodic injection profile surveys can detect formation plugging, injection out of the target zone, thief zones, and underinjected zones. Allocation of injection volumes based upon profile survey results allows tracking of waterflood histories of each zone. Figure 4-18 illustrates the allocation of both injected water and CO_2 relative to the zonal hydrocarbon pore volume of the pattern receiving injection.

Similarly, production profile surveys are run on producing wells to detect formation plugging, production from intended or unintended zones, and zonal communication or non-communication between the injection and production wells. Figure 4-19 illustrates a typical production well profile and also includes snapshots from a downhole video survey run on the well after the production log. In mature production wells, multiple problems are common and can often be identified using the downhole video camera.

Well Pressures

A variety of transient pressure tests can be conducted as follows:[10]

- Pressure buildup (production followed by shut-in) and drawdown (shut-in followed by production) in producing wells

- Pressure falloff (injection followed by shut-in) in injection wells
- Pulse testing involving alternate production and shut-ins of a well giving a series of pressure pulses which are detected at surrounding observation wells

The test results provide reservoir pressure, permeability times thickness, porosity times thickness, skin factor (wellbore damage or improvement), the distance to a fault or other impermeable barrier, reservoir connectivity, and preferential reservoir flow patterns.

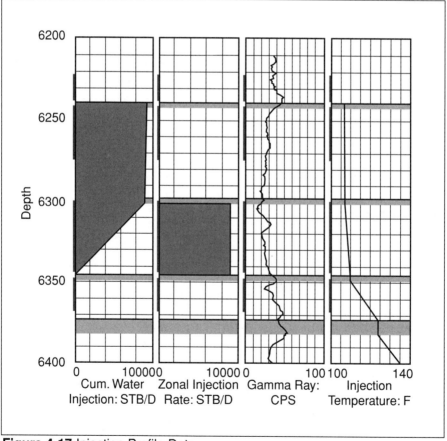

Figure 4-17 Injection Profile Data

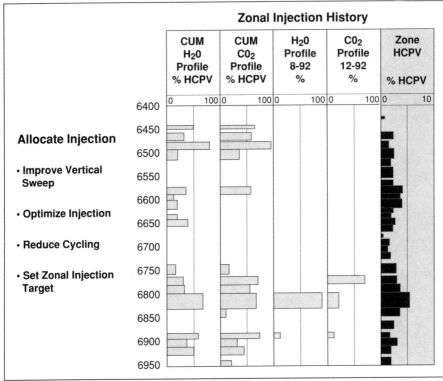

Figure 4-18 Water and Gas Injection Profile Data

Tracer Surveys

Radioactive tracer surveys in injection wells are designed to give permeability profiles. A slug of fluid carrying a soluble, chemically inert, gamma ray emitting isotope is injected into the formation. Gamma ray logs are run prior to and following the tracer injection. Comparison of the pre and post log profiles shows the pay intervals receiving the injected tracers, and thereby the permeability profiles.

Tracer surveys to detect communication between the injector and the surrounding producers can be run by injecting a slug of water containing a soluble tracer followed by water injection. The arrival times of the tracer and its concentration profiles at the producing wells are detected by analyzing the produced water samples. The tracer can be an essentially non- adsorbing chemical or radioactive (tritiated water). The surveys provide reservoir connectivity, and preferential vertical and areal reservoir flow patterns.

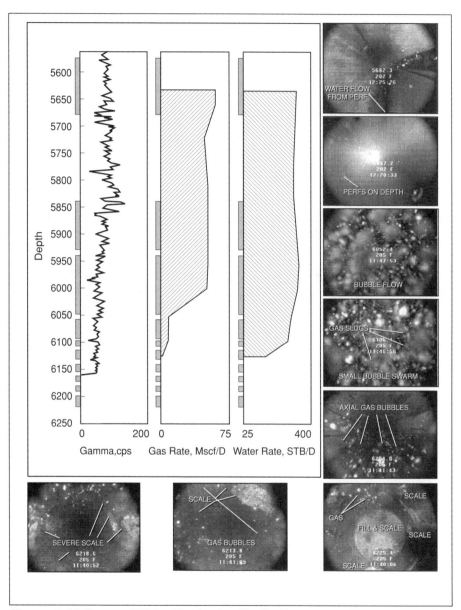

Figure 4-19 Gas and Water Production Profile Data

References

1. Satter, A., and G. C. Thakur. "Integrated Petroleum Reservoir Management: A Team Approach," PennWell Books, Tulsa, OK (1994).
2. A Course in Phase Behavior of Hydrocarbon Reservoir Fluids, Core Laboratories Inc., Houston TX (1990).
3. Dacy, John. Personal communication. Core Laboratories Inc., Houston, TX.
4. Craig, F. F., Jr. "The Reservoir Engineering Aspects of Waterflooding," SPE Monograph 3, Richardson, TX (1971).
5. Amott, E. "Observations Relating to the Wettability of Porous Rocks," Trans., *AIME* (1959) 216, 156-162.
6. Killins, C. R., R. F. Nielsen, and J. C. Calhoun, Jr. "Capillary Desaturation and Imbibition in Rocks," Prod. Monthly (Feb., 1953) 18, No. 2, 30-39.
7. Rose, S. C., J. F. Buckwalter, and R. J. Woodhall. "The Design Engineering Aspects of Waterflooding," SPE Monograph 11, Richardson, TX (1989) .
8. Bilhartz, H. L. "Are We Making Water Systems Too Complex?," Waterflooding, SPE Reprint Series 2, Richardson, TX (1959) 33-36.
9. Hewitt, C. H. "Analytical Techniques for Recognizing Water-Sensitive Reservoir Rocks," *JPT* (Aug. 1963) 813-18.
10. Matthews, C. S. and D. G. Russell. "Pressure Buildup and Flow Tests In Wells," Monograph 1, Richardson, TX (1967).

5

Waterflood Recovery and Factors Influencing Recovery

Introduction

Reservoirs are not uniform in their rock and fluid properties. The variations can be areal as well as vertical. The flood patterns, i.e., injection-production well arrangements, and well completions can be diverse. The overall waterflood recovery efficiency, which is the product of pore-to-pore displacement efficiency and volumetric sweep efficiency, is governed by reservoir heterogeneity, rock and fluid properties, kind and size of flood patterns and well completions.

Appendix D presents a review of reservoir engineering aspects of waterflooding as follows:

- Immiscible displacement theory
- Flood Pattern
- Reservoir Heterogeneity
- Recovery efficiency
- Injection Rates

An in-depth analysis of waterflood recoveries as influenced by important factors listed below were made. The approach to the sensitivity study was to build a 5-spot waterflood computer model for a base case and then vary a specific parameter to analyze the performance. This chapter presents the numerical results of the analysis.

- Timing of waterflood
- Layering attributed to depositional environment and subsequent events (vertical permeability variation)
- Effect of free gas saturation

- Crossflow due to vertical permeability
- Oil gravity

Understanding the effects of these factors on waterflood recoveries is essential in designing, implementing, monitoring, and evaluating waterflood projects.

5-Spot Waterflood Base Case —Primary Depletion

The properties of the base case with Dykstra-Parsons permeability variation factor, V=0.5, are shown in Tables 5-1 and 5-2 and Figure 5-1. Fluid properties and relative permeability data were obtained from correlations (Figures 5-2 to 5-5). The reservoir is under-saturated since original reservoir pressure is above the bubble point, and it is not under natural water drive. The original oil in place was computed to be 843.414 MSTB. The model used the following production limitations:

Economic rate	= 10 STBO per day
Water cut	= 95%
Bottomhole production pressure	= 150 psia
Injection pressure	= 2300 psia

Table 5-1 Properties of the Base Case

Waterflood pattern, acres	40
Total thickness, ft	25
Average permeability, md	24.2
Dykstra-Parsons permeability variation factor (See Figure 5-1 Table 5-2 for, thickness, porosity, permeability distribution)	0.5
Average porosity, %	16.6
Irreducible water saturation, %	22
Critical gas saturation, %	2
Depth, ft	5332
Reservoir temperature, °F	123
Initial reservoir pressure, psia	2332
Bubble point pressure	1855
Oil gravity, °API	33
Gas gravity	0.67 (Air=1)

See Figures 5-2 and 5-3 for oil and gas properties, respectively obtained from correlations
See Figures 5-4 and 5-5 for gas-oil, and water-oil relative permeability, respectively for water-wet sand stone rock

Table 5-2 Porosity-permeability Variation Layers are of Equal 5 ft Thick

	Layer 5	Layer 3	Layer 2	Layer 4	Layer 1
v=0.3					
Probability	10	30	50	70	84.1
Permeability, md	35	27	20	17	14
Porosity, %	17.6	17.1	16.5	16.1	15.7
v=0.5					
Probability	10	30	50	70	84.1
Permeability, md	46	30	20	15	10
*Porosity, %	18.20	17.3	16.5	15.91	5.1
v=0.7					
Probability	10	30	50	70	84.1
*Permeability, md	80	38	20	11	6
Porosity, %	19.4	17.8	16.5	15.2	14.0
v=0.9					
Probability	10	30	50	70	84.1
Permeability, md	280	60	20	6	2
*Porosity, %	21.9	18.8	16.5	14.0	11.7

$$^*\ln(k) = 0.49\phi - 5.09$$

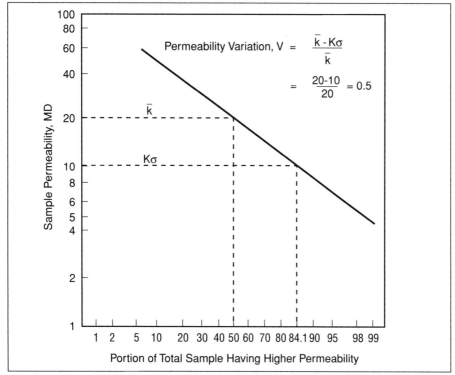

Figure 5-1 Example Problem—Permeability Distribution

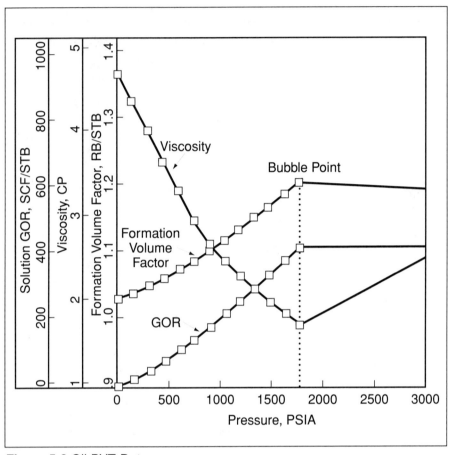

Figure 5-2 Oil PVT Data

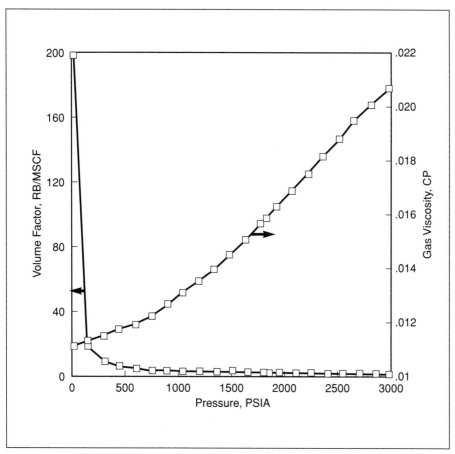

Figure 5-3 Gas PVT Data

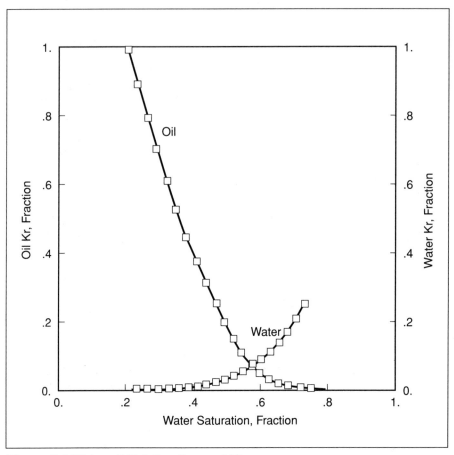

Figure 5-4 Water-oil Relative Permeability

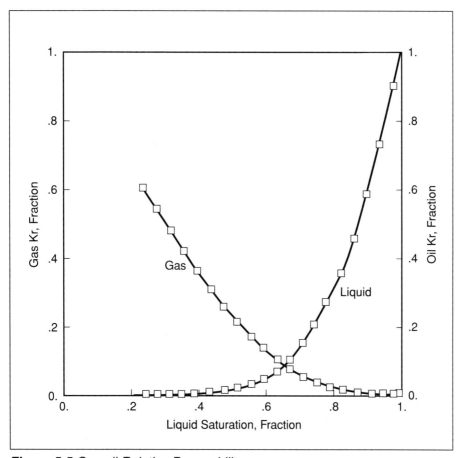

Figure 5-5 Gas-oil Relative Permeability

Primary and waterflood production performance for the five layer 40-acre pattern was simulated by 15x15x5 grid cells model. Figures 5-6 shows pressure and gas-oil ratio vs. recovery efficiency due to primary production.

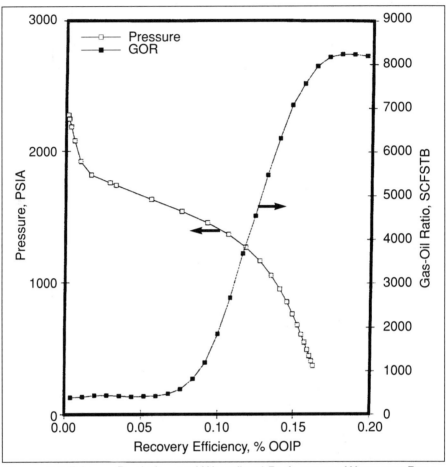

Figure 5-6 Primary Depletion and Waterflood Performance Water-wet Base Case

Timing of Waterflood

Computer runs were made to simulate primary production followed by initiating waterflood at bubble point pressure (P_b), P=1000 psia, and depletion (P = 250 psi). Figure 5-7 shows recovery efficiency vs. pore volume of water injected. Since oil viscosity is at its minimum (see Figure 5-2), total primary and waterflood recovery is significantly more if waterflood is started at bubble point than waiting to waterflood after primary depletion. However, primary recovery is more in case of depletion than earlier waterflooding. Recovery attributed to waterflood is the difference between the total recovery minus the primary recovery.

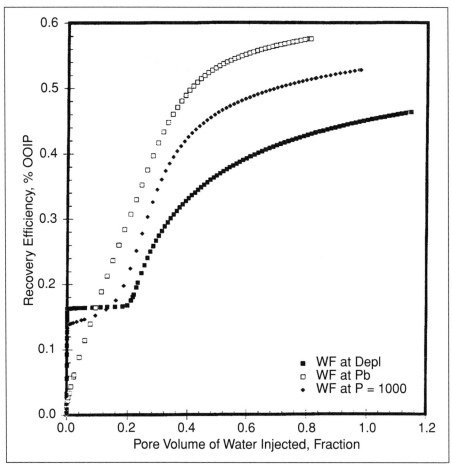

Figure 5-7 Primary and Waterflood Performance After Various Stages of Depletion

The quantity "pore volume of water injected" has an economic significance. For the same recovery, less water injection means shorter project life and favorable economics. Figure 5-7 shows that recovery due to waterflood at bubble point is not only more than flooding at depletion or partial depletion but also economically more beneficial because of less water requirement. Figure 5-8 (water cut vs. pore volume of water injected) shows that water breakthrough occurs earlier in the depletion case than the bubble point case.

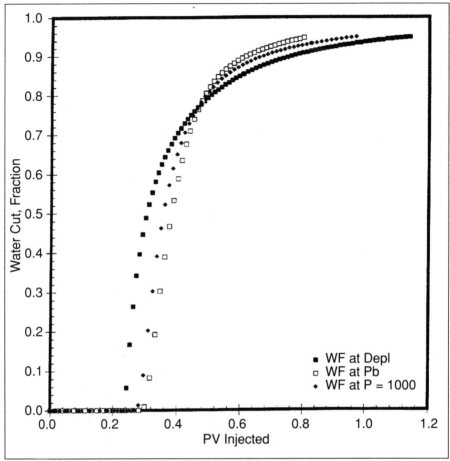

Figure 5-8 Water-cut Performance Primary Followed by Waterflood at Various Stages of Depletion

Layer Permeability Variation

Dykstra-Parsons permeability variation factors, $v = 0$ (homogeneous) to $v = 0.9$ (highly heterogeneous) were used to analyze the layer permeability variation. As the reservoir heterogeneity increases, primary recovery decreases (Figure 5-9) with higher peak gas-oil ratio (Figure 5-10). The heterogeneous cases recover less primary and waterflood oil at the expense of more water injection and earlier water breakthrough (Figures 5-11 and 5-12).

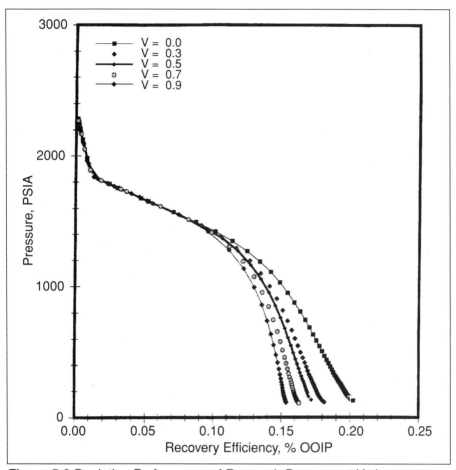

Figure 5-9 Depletion Performance of Reservoir Pressure at Various Dykstra-Parson Ratios

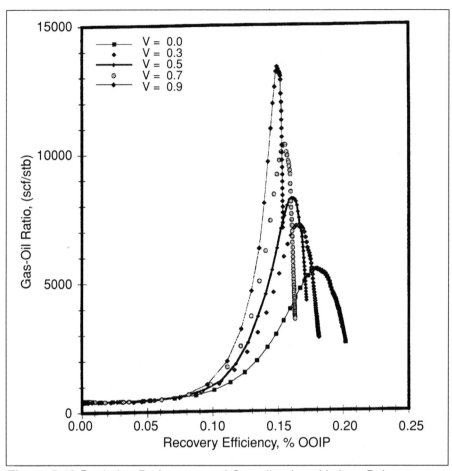

Figure 5-10 Depletion Performance of Gas-oil-ratio at Various Dykstra-Parson Ratios

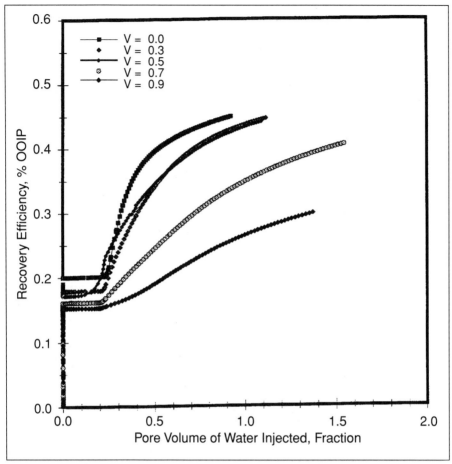

Figure 5-11 Oil Recovery Performance Primary Depletion to 150 psia
Followed by Waterflood at Various Dykstra-Parson Ratios

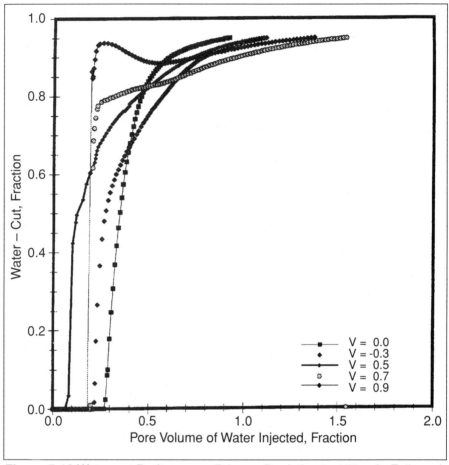

Figure 5-12 Water-cut Performance Primary Depletion to 150 psia Followed by Waterflood at Various Dykstra-Parson Ratios

Critical Gas Saturation

The effects of critical gas saturation on primary and waterflood recoveries were investigated by varying critical gas saturation from 0.02 to 0.10. Higher critical gas saturation yields higher primary oil recovery, higher peak produced gas-oil ratio, and prolonged production at low gas-oil ratio (Figures 5-13 and 5-14). Ultimate primary and waterflood recovery and the amount of water injection are not affected by the variation of critical gas saturation (Figures 5-15 and 5-16).

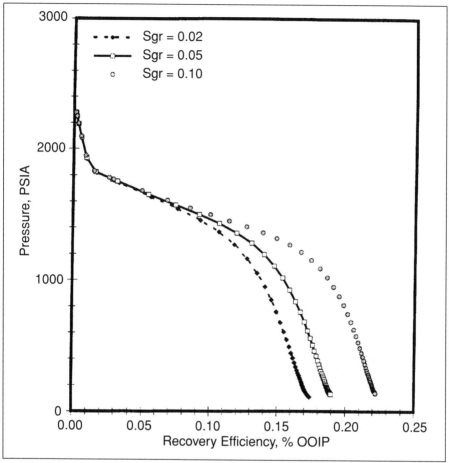

Figure 5-13 Depletion Performance of Reservoir Pressure at V = 0.50 With Various Sgr (Critical Gas Saturation)

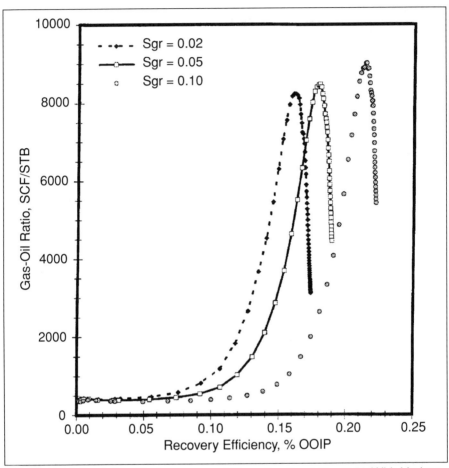

Figure 5-14 Depletion Performance of Gas-oil-ratio at V = 0.50 With Various Sgr (Critical Gas Saturation)

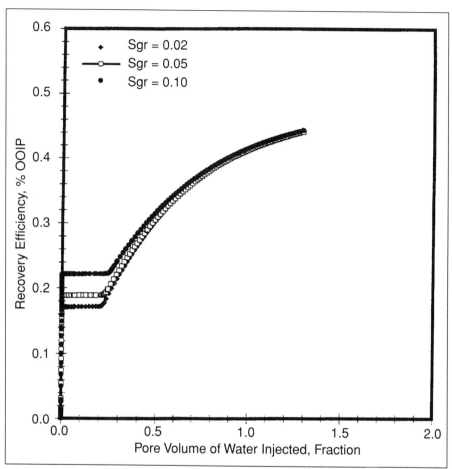

Figure 5-15 Oil Recovery Performance Primary Depletion to 150 psia
Followed by Waterflood at Various Residual Gas Saturations

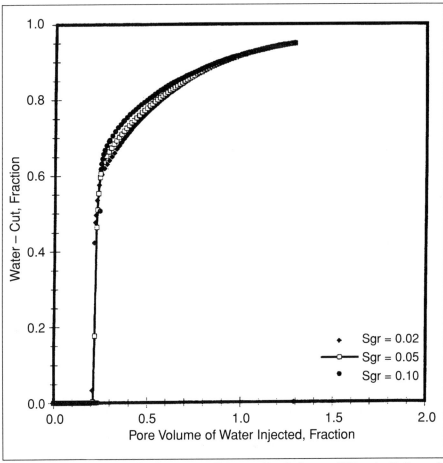

Figure 5-16 Water-cut Performance Primary Depletion to 150 psia Followed by Waterflood at Various Residual Gas Saturations

Vertical Permeability

Crossflow due to vertical to horizontal permeability ratio was studied by varying the ratio from 0.02 to 1.0 (no variation). Primary oil recovery decreases, while the peak produced gas oil ratio increases, with increasing ratio (Figures 5-17 and 5-18). Ultimate primary and water-flood recovery increases with higher ratio (Figures 5-19). Also, the amount of water required is more for the lower ratios to obtain the same oil recovery, and water breakthrough is delayed with higher ratio (Figure 5-20).

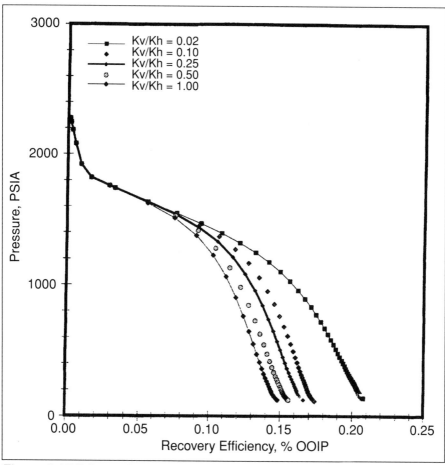

Figure 5-17 Primary Depletion of Reservoir Pressure at V = 0.50 With Various Kv/Kh Permeability Ratios

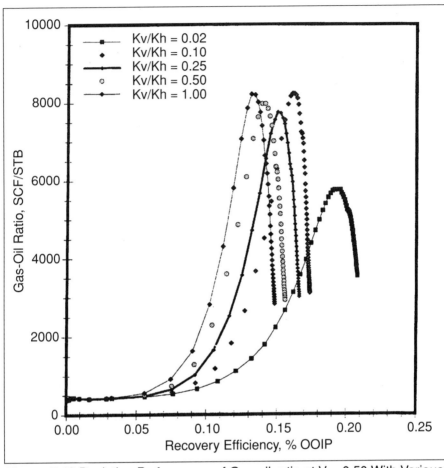

Figure 5-18 Depletion Performance of Gas-oil-ratio at V = 0.50 With Various Kv/Kh Permeability Ratios

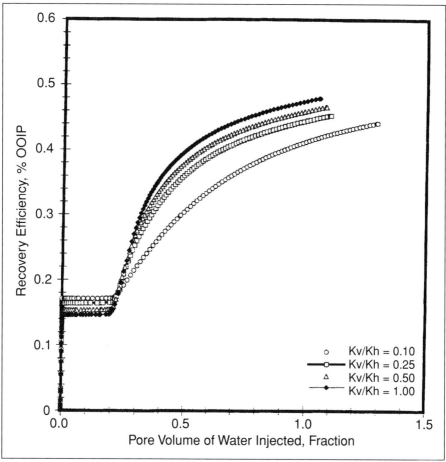

Figure 5-19 Primary Depletion to 150 psia Followed by Waterflood at Various Kv/Kh Ratios - Oil Recovery Performance

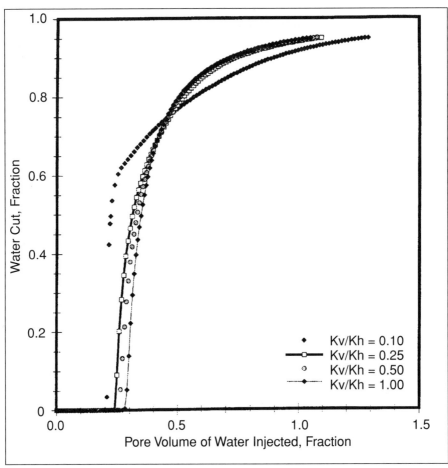

Figure 5-20 Primary Depletion to 150 psia Followed by Waterflood at Various Kv/Kh Ratios - Oil Recovery Performance

Oil Gravity

The effects of oil gravity on primary and waterflood recovery efficiencies were analyzed by varying oil gravity from 10 to 40° API (heavy viscous to light oil). Primary recovery efficiency (Figures 5-21) increases with less viscous oil with higher API gravity. On the other hand, the peak gas-oil ratio (Figure 5-22) is lower in case of higher API gravity oil. Ultimate primary and waterflood recovery increases with higher gravity oil (Figures 5-23). In case of viscous lower gravity oil, water breakthrough occurs earlier than lighter, higher gravity oil (Figures 5-24).

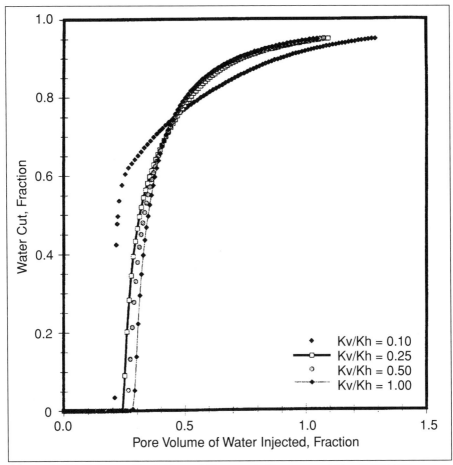

Figure 5-21 Primary Depletion of Reservoir Pressure at V = 0.50 With Various Oil API Gravity

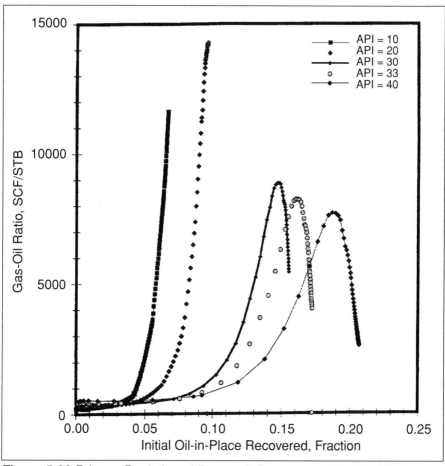

Figure 5-22 Primary Depletion of Reservoir Pressure at V = 0.50 With Various Oil API Gravity

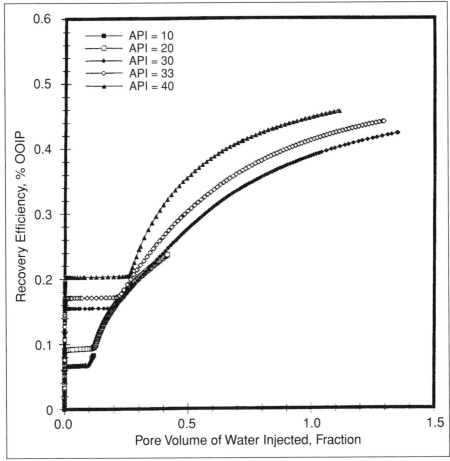

Figure 5-23 Primary Depletion Followed by Waterflood Oil Recovery Performance With Various Oil API Gravity

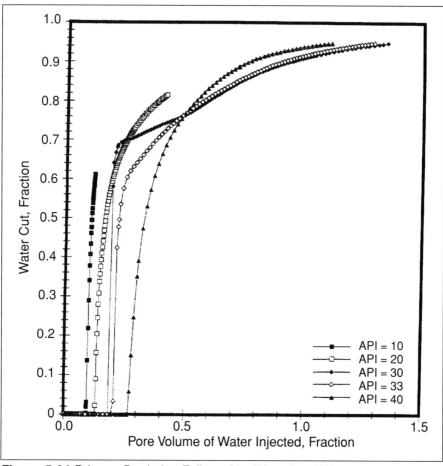

Figure 5-24 Primary Depletion Followed by Waterflood Water-cut Performance With Various Oil API Gravity

6

Infill Drilling

Introduction

Infill drilling (i.e., drilling of additional wells after initial primary and/or secondary development of a field) played an important role in improving waterflood recovery in West Texas in the 1970s. Infill drilling would essentially result in acceleration of production in an ideal homogeneous reservoir but no incremental recovery. However, additional recovery can be obtained in heterogeneous reservoir without reservoir continuity.

Figure 6-1 shows a plot of oil production rate vs. time for decline curve prior to infill drilling (base case), acceleration, and incremental recovery cases. In case of acceleration, the area under the acceleration curve is equal to the area under the decline curve. Also, the areas A and B between the acceleration and decline curves are equal. In case of incremental recovery, the area under the incremental recovery curve is greater than the area under the decline curve. Also, the area C between the incremental recovery and decline curves is greater than the area D between the incremental recovery and decline curves.

Figure 6-2 shows a schematic of discontinuous layered reservoirs with injectors and producers. The layer connectivity and waterflood recovery potential are given in Table 6-1. Ideally speaking for optimum recovery, the injectors and producers need to penetrate all the oil bearing layers which require closer spacing.

Table 6-1 Connectivity and Waterflood Recovery Potential

Layer	Connectivity	Waterflood Recovery
1 and 3	Injection well with both producers	From both producers
2	Injection well with producer 1	From producer 1 only
4	None	None
5	Injection well with producer 1	From producer 1 only
6	Injection well with producer 2	From producer 2 only

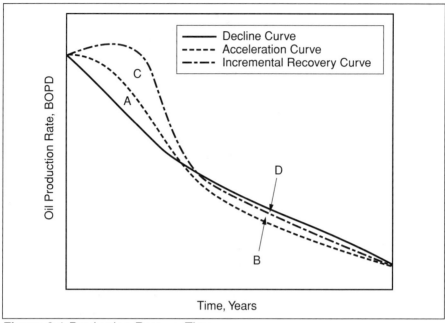

Figure 6-1 Production Rate vs. Time

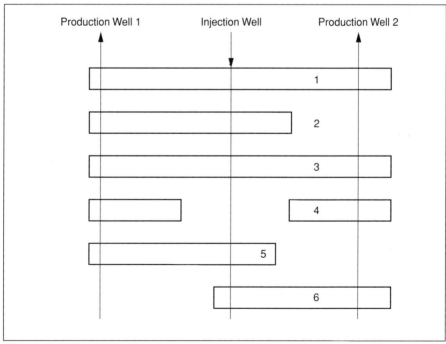

Figure 6-2 Structural Cross Section

Infill drilling is now recognized as a viable improved recovery process in primary, secondary, and tertiary operations.. This chapter discusses the importance of infill drilling and how to select infill wells (e.g., on what spacing) in a waterflood operation. In addition, the benefits of infill drilling will be illustrated with the help of case examples. These cases will discuss geological and geostatistical models with continuous and discontinuous pay zones and will explain how infill drilling can add reserves by improving flow continuity or continuity between the injectors and producers.

Why Infill Drilling ?

It is not uncommon to start a waterflood using a peripheral flood, centerline flood, or a combination of these in order to minimize number of injection wells and capital investments. However, most of these floods take a long time to increase oil production as a result of large distances between the injectors and the last row of producers and especially if the formation permeability is low. Also, in some floods, it becomes very difficult to maintain high enough reservoir pressure because of a significantly higher number of injection wells required relative to the producing wells in order to balance injection and production rates.

Pattern waterfloods can provide pressure support by maintaining balanced injection and production volumes and also early oil production response. The reservoir heterogeneity and layer discontinuity or connectivity can be controlled via infill drilling, affecting the well spacing. The infill drilling reducing the well spacing enhances the injection/production well connectivity. Wu, et al.[1] reported the results of a study to determine the impact of infill drilling on the waterflood recovery in West Texas carbonate reservoirs. Figure 6-3, a plot of waterflood recovery vs. well spacing, based upon an analysis of the performances of 24 reservoirs, shows a certain degree of correlation trend between the waterflood recovery and the well spacing.

Goodwin presented a case history of infill drilling in the Means San Andres Unit in West Texas.[2] More than 500 infill wells had been drilled as the unit had gone from primary on 40 acre spacing, through waterflooding on 20 acre spacing, to CO_2 miscible flooding on 10 acre spacing.[2] He concluded that infill drilling and flood pattern modification programs added significantly to the ultimate recovery from this field. Oil reserves had been increased over 40% by the implementation of these programs.

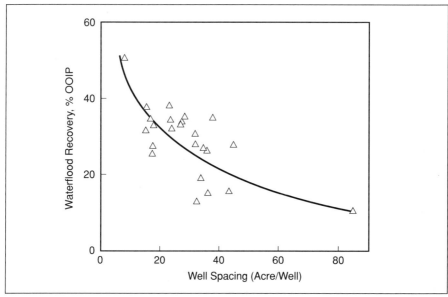

Figure 6-3 Waterflood Recovery vs. Well Spacing (Wu, C.H., et al. "An Evaluation of Waterflood Infill Drilling in West Texas Clearfork and San Andres Carbonate Reserves," SPE Paper 19783 presented at the SPE Annual Technical Conference and Exhibition in San Antonio, TX, Oct. 8-11, 1989)

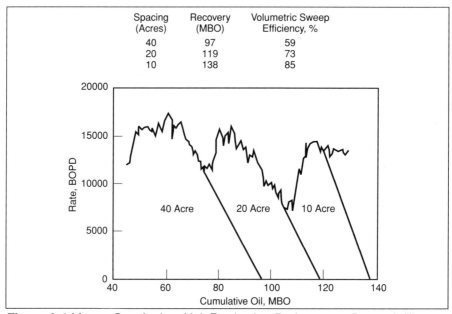

Figure 6-4 Means San Andres Unit Production Performance Due to Infill Drilling (Goodwin, J.M. "Infill Drilling and Pattern Modification in the Mean San Andres Unit.")

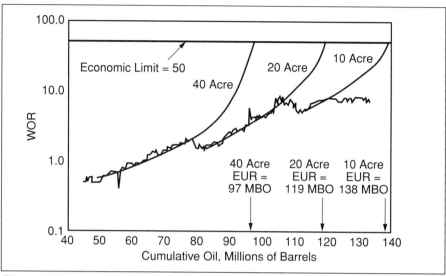

Figure 6-5 Means San Andres Unit Water-cut Behavior Due to Infill Drilling (Goodwin, J.M. "Infill Drilling and Pattern Modification in the Mean San Andres Unit.")

Figures 6-4 and 6-5 show Means San Andres Unit production performance, and water cut behavior due to infill drilling, respectively. Volumetric sweep efficiency had been increased by increasing interwell continuity through closer well spacing i.e., 59% in case of 40 acre spacing to 85% for 10 acre spacing. Flood conformance had been improved by rotation of the flood pattern. Figure 6-6 shows modification of flood patterns from 40 to 10 acres. Crossflooding of the pattern by new injectors had contacted new, undisplaced oil as shown schematically in Figure 6-7.

The factors contributing to increased recovery after infill drilling are listed below:[3,4,5]

1. Improved areal sweep
2. Improved injection imbalance due to areal heterogeneity
3. Improved vertical sweep
4. Lateral pay continuity
5. Recovery by wedge-edge oil
6. Reduced economic limit

In improving areal sweep efficiency, oil that is held up in the corners is immediately swept by reversing the streamlines within the pat-

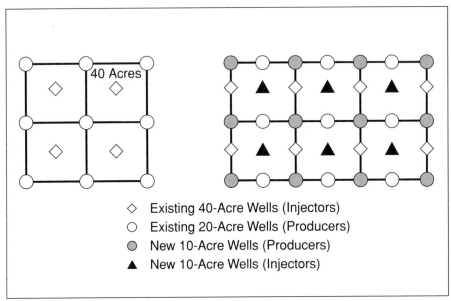

Figure 6-6 Means San Andres Unit Flood Pattern (Goodwin, J.M. "Infill Drilling and Pattern Modification in the Mean San Andres Unit.")

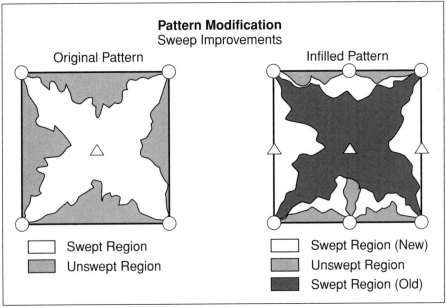

Figure 6-7 Means San Andres Unit, Sweep Improvements Due to Pattern Modification With Infill Drilling (Goodwin, J.M. "Infill Drilling and Pattern Modification in the Mean San Andres Unit.")

tern. Also, patterns which are subjected to poor geometric alignment and resulting poor streamline balance, can be significantly improved with additional well locations. Figure 6-8 shows a schematic of simulated streamlines and oil saturations before and after infill drilling.

In a pattern waterflood, infill drilling can improve injection imbalance due to areal heterogeneity which causes early water breakthrough and preferential sweep of only part of the pattern. Vertical sweep can be also improved by converting producers to injectors and drilling new producers with selective perforation to isolate thief zones.

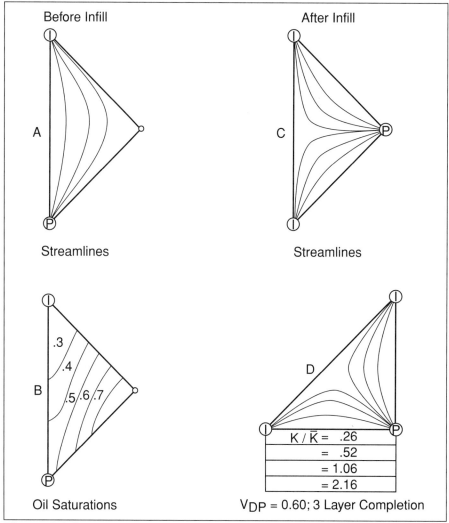

Figure 6-8 Schematic of Infill Cases

As has been already discussed, lateral pay connectivity can be achieved by infill drilling contributing to additional recovery. The infill well between two original wells opens up new pay which did not exist between the original wells, leading to incremental oil recovery.

Wedge–edge oil results from formation dip at oil-water contacts or inadequate edge development of the pattern area. Infilling with small patterns can result in flooding of previously unflooded area or zones.

Acceleration of recovery is one of the economic benefits of infill drilling. In addition to having more producers, the injection rate is increased more than the well ratio might indicate. This can happen because the pressure drop between the injector and producer occurs over a shorter distance. Also, operating costs are reduced as the water cut is significantly decreased, and the economic limit for a project will be improved as a result of the operating changes.

Selection of Infill Wells

Several factors must be considered in selecting an infill project as follows:

- Production/injection performance
- Reservoir description
- Infill drilling project design
- Economic evaluation

Analyze water breakthrough and oil recovery relative to expected theoretical oil recovery based upon layered reservoir, mobility ratio, and pattern type. In general, the worse the original waterflood efficiency, the better is the infill opportunity. In an ideal flood, all injectors and the injection/production ratio should be balanced. If it is not reasonably balanced, infill drilling and/or pattern balancing need to be considered.

An integrated reservoir model (Figure 6-9) which will require a thorough knowledge of the geology, geophysics, rock, and fluid properties, fluid flow and recovery mechanisms, drilling and well completions and past production performance will play an important role in developing an appropriate strategy for infill drilling. Simulation of the past full field or pattern production performance based upon an integrated reservoir description will provide a valuable clue to design a plan for infill drilling.

A project design phase should be performed with a good reservoir description to determine the expected infill well performance for each pattern and to make economic evaluation.

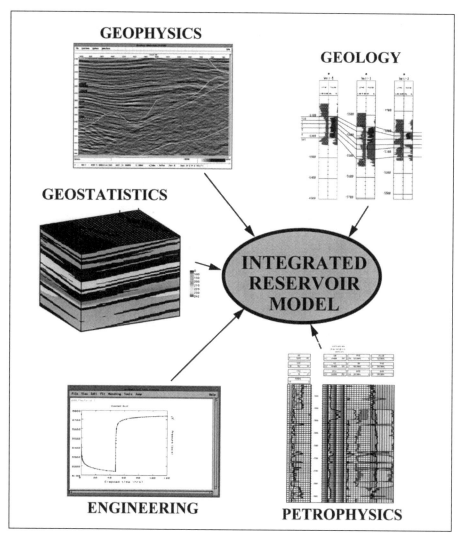

Figure 6-9 Integrated Reservoir Model

Reducing pattern size is not necessarily the best strategy for infill drilling. Akinlawon, et al.[6] reported how locations of two horizontal wells in North Apoi/Funiwa Field in Nigeria were chosen based upon simulated oil saturation distributions. An integrated reservoir optimization study of the field was carried out in 1995 by a team of geoscientists and engineers in order to evaluate and capture the upside potential of this offshore mature field. The study included reviewing and ver-

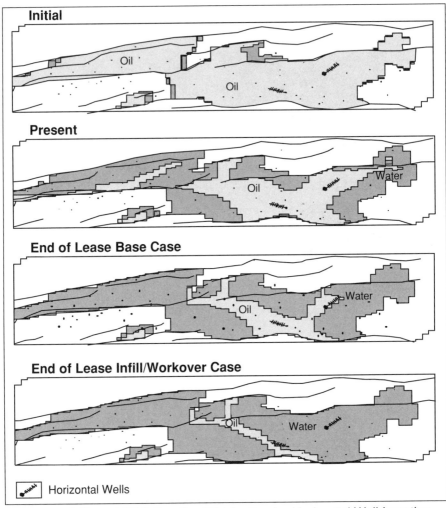

Figure 6-10 Use of Oil Saturation Distribution for Horizontal Well Location

ifying available geoscience and engineering data, building a reservoir characterization model, and matching past production history to validate the model. The resulting mathematical model of the reservoir was then used to predict future reservoir performance under various investment scenarios for optimally draining the reservoirs, including additional take points, horizontal wells, gas lift and water injection.

The optimum locations of the two infill horizontal wells recommended in the phase 1 study were determined from computed residual oil saturation distributions (Figure 6-10). The first well came on production at 2,670 BOPD from a 700-foot horizontal section. The second well had a 1600 foot horizontal pay section and started producing at 4,020 BOPD. Both wells performed according to plan and together they are now producing more than 6,600 barrels of oil per day. This example demonstrates how computed oil saturation distribution can be utilized with confidence to select the location of infill wells.

Although other ranges of properties are applicable for infill drilling, Reviere and Wu[7] reported the following ranges of reservoir properties for favorable infill drilling based upon an economic evaluation of nine Texas waterflood units.

- Depth, ft   4,300 to 7,000
- Net pay, ft   12 to 500
- Permeability, md   0.7 to 27
- Porosity, %   7 to 19
- Water saturation, %   20 to 45

Determining the incremental oil from infill drilling is the major challenge to infill well design and implementation. Lu, et al.[8] presented empirical equations relating infill drilling ultimate recovery with the estimated oil-in-place per well at the start of infill drilling, average annual rate of incremental oil production from infill drilling, the well spacing reduction, the net thickness, and the reservoir depth. The correlations were based upon production performance and economic evaluation of 21 San Andres carbonate units.

Chan, et al. presented techniques for estimating recovery due to infill drilling as described below: [9]

- A reservoir continuity model illustrating the concept that the infill drilling enhances the continuity between the wells, thereby improving the sweep efficiency and ultimate recovery
- A plot of log of water oil ratio vs. cumulative oil production to give the incremental oil recovered from infill drilling
- Decline curve analysis to estimate the incremental recovery and accelerated production, and the interference between infill and the offset wells
- A reservoir simulation model to predict production forecast for the infill wells, and to estimate the incremental recovery vs. the accelerated production from infill drilling

The authors warned that no one method is necessarily more accurate than another. However, more confidence would be established by using all four methods.

The U. S. Department of Energy created an infill drilling predictive model (IDPM) for waterflood infill behavior. The model is a hybrid between streamtube and three-dimensional two-phase numerical simulator. Fuller, et al. used this model to determine its usefulness to screen for infill drilling opportunities. [10]

Lens Model reported by Christman, which is based upon a physical concept of the individual pay, is designed to calculate reserves from infill drilling and pattern modification. As derived, the model can be applied to 5-spot and line drive patterns. [11] It was calibrated against the performance of several Wasson Clearfork Trend waterfloods by estimating the primary and secondary recoveries and adjusting the lens size distribution.

Case Examples

Barber, et al. presented actual infill drilling performance of 9 fields in West Texas, Oklahoma, and Illinois, including Means, Fullerton, Robertson, IAB (Menielle Penn), Howard Classcock, Dorward, Sand Hills in West Texas, Hewitt in Southern Oklahoma, and Loudon in Illinois.[12] They include dolomite, limestone, and sandstone reservoirs with porosities varying from 4 to 21% and with average permeabilities varying from 0.65 to 184 md. The results showed that additional oil recovery was realized by improving reservoir continuity with increased well density.

Reviere and Wu[7] presented results of their economic evaluation of infill drilling in 9 Texas waterflood units. The fields include Fullerton, Big Wells (San Miguel), Robertson (Clearfork), IAB (Menielle Penn), Means (San Andres), Wasson (San Andres), Fuhrman Mascho Black 9, and Triple-N (Grayburg). The actual oil production performances were compared with predicted results by decline curve analysis. The results indicate that waterflood infill drilling produced substantial incremental oil and the economic performances can be characterized from good to excellent.

References

1. Wu, C. H., et al. "An Evaluation of Waterflood Infill Drilling in West Texas Clearfork and San Andres Carbonate Reserves," SPE Paper 19783 presented at the SPE Annual Technical Conference and Exhibition in San Antonio, TX., Oct. 8-11, 1989.
2. Goodwin, J. M. "Infill Drilling And Pattern Modification In The Means San Andres Unit."
3. Driscoll, V. J. "Recovery Optimization Through Infill Drilling – Concepts, Analysis, Field Results," SPE Paper 4977 (Oct. 1974).
4. Gould, T. L. and M. A. Munoz. "An Analysis of Infill Drilling," SPE Paper 11021 presented at the SPE Annual Technical Conference and Exhibition in New Orleans LA, Sept. 26-29, 1982.
5. Gould T. L. and A. M. S. Sarem. "Infill Drilling for Incremental Recovery," JPT (March 1989) p 229-237.
6. Akinlawon, Y., T. Nwosu, A. Satter, and R. Jespersen. "Integrated Reservoir Management Doubles Nigerian Field Reserves," Harts Pet. Eng. Int'l, Oct. 1996.
7. Reviere, R. H. and C. H. Wu. "An Economic Evaluation of Waterflood Infill Drilling in 9 Texas Waterflood Units," SPE Paper 15037 presented at the SPE Permian Basin Oil and Gas Recovery Conference in Midland, TX., March 13-14, 1986.
8. Lu, G. F., B. K. Jagoe, and C. H. Wu. "Technical Factors useful for Screening Carbonate Reservoirs for Waterflood Infill Drilling," SPE Paper 27660 presented at the SPE Permian Basin Oil and Gas Recovery Conference in Midland, TX, March 16-18, 1994.
9. Chan M. C. F., S. J. Springer, S. Asgarpour, and D. J. Corns. "Evaluation of Incremental Recovery By Infill Drilling." Paper No. 86-37-14 presented at the 37th Annual Technical Meeting of the Pet. Soc. Of CIM in Calgary, June 8-11, 1986.
10. Fuller, S. M., A. M. Sarem, T. L. Gould. "Screening Waterfloods for Infill Drilling Opportunities," SPE Paper 22333, presented at the SPE International Meeting of Petroleum Engineering in Beijing, China, March 24-27, 1992.
11. Christman, P. G. "Modeling the Effect of Infill Drilling and Pattern Modification in Discontinuous Reservoirs," SPE/DOE 27747, presented at the SPE/DOE Symposium on Improved Recovery in Tulsa, Oklahoma, April 17-20, 1994.
12. Barber, A. H., Jr., C. J. George, L. H. Stiles, and B. B. Thompson. "Infill Drilling To Increase Reserves – Actual Experience in Nine Fields in Texas, Oklahoma, and Illinois," JPT (Aug. 1983).

7

Design of Waterfloods: Geological, Engineering and Operational Aspects

This chapter discusses the design of waterfloods from an integrated point of view, involving geological, engineering and operational aspects. A five-phase, well organized waterflood design process is described. In addition, detailed project design considerations, items of interest related to reservoir characterization, process and operation design, equipment design, data acquisition design, economic evaluation (including project scale-up), why waterfloods fail, and an example of waterflood design are presented.

Design and Operation Process

The following five phases describe the design and operation process for waterflooding (see Figure 1-2).

Phase 1 – Broad, Conceptual Design to Identify and Frame Opportunity

The first step in conceptual design is to identify the business opportunities (declining reserves, emphasis on replacing or increasing reserves, reservoir performance under primary depletion, successful waterfloods in the same or similar reservoirs). The next step is to form a task force or a team to perform a quick waterflood feasibility study to add probable reserves. The study should include an integrated approach, involving various pertinent functions, and develop:

- Location of field (onshore/offshore)
- Reservoir characteristics, depth, thickness, temperature, oil gravity and viscosity
- Probable injection pattern and alternatives
- Approximate values of rates and pressure (injection and production fluids)
- Water source, quality, compatibility
- Project life
- Assimilate information available for the project, reservoirs, wells and facilities (as shown by an integrated waterflood system in Figure 1-2). Also document what is not available
- Rough cost estimates for pump system, water source and treating, modification of existing facilities, drilling and completing well, changing well equipment, operations
- Coarse economic screening to determine economic feasibility using reasonable estimates for parameters

The result of this phase is to determine the potential economic attractiveness of the project.

Phase 2 – Generate and Select Alternatives

The second phase steps are:

- Collect, select, validate, consolidate, manage and store information for reservoir study and asset development planning (see Tables 7-1 and 7-2)
- Multifunction team to perform a study and look at several development alternatives (for more detailed study) generating the following:
  - reservoir description
  - drilling and production performance analysis
  - waterflood performance prediction utilizing reservoir simulation and/or classical methods
  - waterflood analysis of similar fields, if available
  - overall integrated waterflood system
  - economic analysis and risk assessment, clearly identifying economic drivers and destroyers
  - initiate unitization proceedings, involving other partners
  - plan a pilot study and/or injectivity test, if necessary

Table 7-1 Step-by-step Methodology (Thakur, G.C. "Reservoir Management of Mature Fields," IHRDC Video Library for and Production Specialists, 1992)

1. Review previous studies
2. Review field development and performance, including primary, secondary, and enhanced recovery operations, paying particular attention to
   - drive mechanisms
   - production and pressure history (deduce past performance)
   - recovery factors
   - mobility ratio and sweep efficiency
   - well spacing and drainage
   - wells and facilities conditions
   - lease line migration
   - IPR curves/individual well performance
3. Establish geological parameters, including
   - general reservoir configuration
   - fluids distribution and movement in the reservoir
   - variation in pore space properties
   - continuity and thickness
   - porosity
   - thickness and structure maps
   - rock type, porosity and permeability cutoffs
   - fluid contacts
   - 3-D description of reservoir
   - vertical stratification
4. Determine "pay." Compare log and core data with DST (Drill Stem Test) and RFT (Repeat Formation Test) data.
5. Estimate OOIP (original oil in place) and OGIP (original gas in place)
6. Review database and monitoring program, including
   - logging, well testing
   - injection profiles
   - completion/workover records
   - PVT analysis
   - relative permeability, capillary pressure, and coreflood and wettability tests
   - surface facilities data
   - pattern performance monitoring (areal flood balance and vertical conformance monitoring)
   - interwell tracers
   - well performance
7. Use simulators to perform history matching
8. Estimate reserves (original and remaining) and forecast production
   - geology and formation evaluation data
   - drive mechanisms
   - fluid properties
   - relative permeability and residual saturation data
   - reserve

Table 7-2 Types of Information

Field Information:
        Physical description of the reservoir
        Surrounding environment
Geologic and Engineering data:
        Physical boundaries
        Reservoir characteristics:
                Pay quality and continuity
                Zonation and Heterogeneous effects
                Permeability direction
                Fracture orientation
                Unusual completions
Primary Operations History:
        Primary recovery data
        Production equipment installed
        Well-completion data
Waterflood Layout:
        Potential pattern selection
        Selection of wells and spacing
        Production loss from injection wells
        Possible expansion
        Terrain and topography
Pilot Tests

Phase 3 – Develop Preferred Alternatives

The third phase involves performing a detailed evaluation of preferred waterflood alternatives (e.g., 5-spot, 9-spot, line drive, peripheral flood, etc.), utilizing core and log information, completion data, production performance analysis (oil, water, GOR), pressure performance analysis, reservoir fluid analysis, operating costs, laboratory and pilot test results, etc. Successfully managing waterflood projects from inception to abandonment requires experience and technical expertise in a variety of disciplines. Contributions from drilling engineers, offshore development project managers, geologists, reservoir engineers, and environmental protection specialists, among others, are essential for economically successful waterfloods.

This job aid summarizes key concepts in the design and management of a waterflood. It was designed to help you do your work better and includes the following information:

- Illustration of typical flood patterns
- Checklist for estimating waterflood performance

- Process for waterflood management
- Screening criteria for evaluating empirical performance factors
- Guidelines for calculating oil saturation
- Table of potential waterflood problems with probable causes and solutions

See Table 7-3 for a detailed description of waterflood design, including evaluation of the reservoir, selecting potential flooding plans, estimating injection/production rates, forecasting oil recovery, designing

Table 7-3 Flood Patterns and Waterflood Design

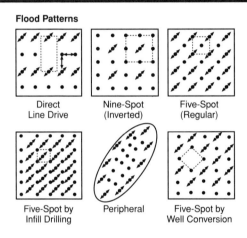

Flood Patterns

Direct Line Drive   Nine-Spot (Inverted)   Five-Spot (Regular)

Five-Spot by Infill Drilling   Peripheral   Five-Spot by Well Conversion

1. Evaluate the Reservoir
   - Reservoir Characterization
   - Formation Evaluation
2. Select Potential Flooding Plans
   - Peripheral Flood
   - Pattern Configuration
   - Aquifer Injection
   - Well Spacing
3. Estimate Injection/Production Rates
   - Injectivity Tests
   - Empirical Correlations (Rules of Thumb)
   - Local Experience
4. Forecast Oil Recovery Over the Life of the Project for Each Flooding Plan
   - Material Balance
   - Empirical Correlations
   - Analytical Models
   - Reservoir Simulators

5. Preliminary Facilities Design
   - Estimate Fluid Volumes and Rates for Sizing Equipment and Fluid Handling Systems
   - Identify Compatible Water Source or Injection
   - Arrange for Disposal of Produced Water
6. Estimate Capital Expenditures and Future Operating Expenses
   - Facilities
   - Wells
   - Lifting Costs
   - Treating Costs Handling
7. Conduct Decision Analysis and Economic Evaluation
8. Identify Variables That May Cause Uncertainty
   - Original Oil-in-Place
   - Sweep Efficiency
   - Injection Rates
   - Reservoir Discontinuities

facilities, estimating capital expenditures and operating expenses, conducting decision analysis and economic evaluation, and identifying variables that may cause uncertainty.

- Select best alternative(s) for detailed design and development, and develop detailed waterflood design plans and estimate development costs
- Perform detailed risk and environmental assessments
- Negotiate unitization parameters, if necessary
- Perform economic and risk analysis, and develop an expected value of the project
- Seek management approval and government approval
- Develop and document waterflood asset management plan, including metrics

Phase 4 – Implement Waterflood and Its Management Plan

- Complete detailed design, sizing and selection of materials and equipment
- Implement asset management plan
  - drill and/or recomplete wells
  - build facilities
  - install field and artificial lift equipment
- Finalize waterflood operating plan and schedule
- Develop waterflood surveillance and monitoring plan
- Select operating team and train

Phase 5 – Operate, Monitor and Evaluate Waterflood

- Monitoring reservoir, wells and facilities performance
- Evaluate performance against metrics
- Modify "living" reservoir model as additional data are obtained from operational and evaluation of results
- Revision of plan and strategies based upon actual performance
- Identify new opportunities for expansion
- Plan exit strategy for terminating the waterflood at some point in time

Table 7-4 describes some potential problems, considerations/causes, and possible remedies/solutions.

Table 7-4 Potential Problems

Reservoir	Considerations/Causes	Remedies and Solutions
❏ Low Displacement Efficiency	• Permeability Anisotropy, Formation Fractures, Adverse Mobility Ratio, Formation Dip	• Pattern Balancing, Pattern Realignment, Recompletions, Infill Wells
❏ Channeling (Thief Zones)	• High Permeability Streaks	• Chemical/Mechanical Isolation
❏ High Gas Saturation	• Delay in Implementating Water Injection	• Shutting in High-Gas Producing Wells • Injection Greater than Production • Proper Selection of Production and Injection Wells
❏ High Water Saturation	• Aquifer Water Influx	• Avoid High Water Saturation Area: Use Polymers and Chemicals to Reduce Water Cycling
❏ Unbalanced Injection Pattern	• Evaluation Oversight, Permeability Anisotropy, Injectivity Problems	• Evaluate and Understand Reservoir Characterization and Pattern Balancing, Make Necessary Adjustments

Production Wells	Considerations/Causes	Remedies and Solutions
❏ Scale Deposition (Calcium Carbonate or Barium Sulfate)	• Dissimilar Formation Water, Chemical Reactions	• Acid, Mechanical, Chemical Solvents, Chemical Treatment Program
❏ Acid Sludge, Iron Sulfide	• Chemical Reactions	• Chemical Solvents
❏ Formation Damage	• Asphaltenes, Paraffins, Acid Sludge, Calcium Carbonate	• Chemical Solvents, Inhibitors, Acid/ Anti-Sludge Additives, Mechanical
❏ Channeling (Thief Zones)	• High Permeability Streaks, Natural Fractures	• Chemical/Mechanical Isolation, Pattern Realignment
❏ Excessive Pump Failures/Rod Parts	• Inadequate Fluid Levels, Poor Pumping System Design	• Pump-Off Controller, Modify Pumping System

Injection Wells	Considerations/Causes	Remedies and Solutions
❏ Injectivity Loss	• Formation Damage, Increasing Reservoir Pressure	• Isolated Fractures, Stimulation
❏ Conformance	• High Permeability Streaks, Fractures, Differences in Formation Pressure	• Chemical/Mechanical Isolation
❏ Formation Damage	• Fines Migration, Asphaltenes, Paraffins, Iron Sulfides	• Acid, Chemical Solvents, Inhibitors, Isolated Fractures
❏ Oil Carry-Over	• Improper Design or Plugging of Free Water Knockout	• Maintain Free Water Knockout
❏ Profile Modification	• Difficult to Apply	• Focus on Quality Control of Chemicals and Workover Procedure

Facilities	Considerations/Causes	Remedies and Solutions
❏ Limited Injection Capacity	• Improper Facilities Design	• Good Communication Between Disciplines Involved in Project
❏ Scale Deposition, Corrosion	• Dissimilar Formation Water	• Inhibitors, Separate Dissimilar Water
❏ Poor Water Quality	• Filtering System, Free Water Knockout	• Maintain Filter System and Free Water Knockout

Project Design Considerations

Whether we are planning a full-field project or a small scale evaluation project, the overall design and integration of each aspect of the project is critical to its success. In a full-field project, the measure of success is strictly economic. In a small-scale evaluation or pilot project, success lies in acquiring the data needed to provide a meaningful interpretation of process performance, regardless of the project's economic benefits. The various stages, or areas, of project design discussed below are applicable to both full-scale and pilot projects. The only differences are the degree of application involved and the fact that a pilot project allows room for mistakes and an opportunity to improve the process at minimal cost, while the same mistakes made in a full scale project could spell economic disaster.

We have already stated that the purpose of a pilot is to provide a meaningful interpretation of the performance of a given enhanced recovery process (regardless of the project's economic success) by acquiring data under representative field conditions. Our understanding of the process and, subsequently, our ability to confidently expand the project to full-scale conditions depends on our being able to predict the performance of the pilot.

To further clarify its purpose, a pilot test is

- an experiment in which the field is the laboratory
- a simulation of the larger field effort
- a place to make your mistakes—before they get too costly
- a place to work the bugs out of equipment
- a place to develop the controls and data needed to insure the success of a project
- a place to put your best engineering talent to the test, and to train other engineers
- a place for intensive planning
- a place for open minds and compromise
- a place for finding every problem imaginable (and many not yet imagined), with their attendant delays

A pilot test is not

- a short-term money-making proposition
- a total field simulation
- a great place (usually) to demonstrate "theory in action"

Reservoir Characterization

To predict the performance of a waterflood we must know the characteristics of the reservoir in the test area or field. For our present purposes, these characteristics include both reservoir fluid and reservoir rock parameters. However, in addition to rock and fluid properties, there are more general parameters that are critical to the successful interpretation of all waterfloods. They include:

- Oil saturation
- Water and gas saturations
- Fluid transport properties and phase equilibrium data
- Local rock properties (permeability, porosity and thickness)
- Regional rock properties
- Regional fluid distributions and fluid dynamic conditions (water zones or aquifers, gas caps, fluid pressure and distribution, fluid movement, etc.)

Let's examine the relevance or importance of these parameters to the successful completion of a waterflood process, and briefly discuss how they can be determined.

Oil Saturation

Knowing the volume of oil present in a reservoir at the start of a project is critical in evaluating the economics of any waterflooding project. Knowing the initial oil saturation is critical for a secondary recovery process which typically requires a heavy front end investment. Therefore, a small error in estimated oil saturation (as small as 5% pore volume) can significantly affect the economics of the project. The oil saturation also affects the life of the project. The amount of time required to bank the oil and drive it to a producing well also affects the profitability of a waterflood.

Thus, because of its large impact on project feasibility, the current oil saturation determination is critical. It is, however, a difficult parameter to measure in real reservoirs.

Several methods exist to determine residual/remaining oil saturation, including:

- Analysis of either conventionally or specially obtained core samples
- Reservoir engineering calculations, which range from simple material balance methods to detailed numerical simulation studies of past recovery processes
- Well logging of various types (See Table 7-5 for a listing of various applicable types)
- Pressure transient measurement, applied alone or in combi-

nation with core analysis and/or experimental relative permeability data
- Chemical tracers studies

Table 7-5 Basic Tools and Techniques to Determine Residual Oil Saturation (Thakur, G.C. "Reservoir Management of Mature Fields," IHRDC Video Library for Exploration and Production Specialists, 1992)

Basic Tool	Technique	Can Be Used When Hole is Cased	Has Been Field Tested	Expected Accuracy[1]
Reservoir Performance	Volumetric determination	Yes	Yes	Poor
	Pressure Transients			
Well Tests	Fluid Compressibility	Yes	Yes	Very Poor
	Effective Permeability	Yes	Yes	Poor
	Water/Oil Ratio	Yes	Yes	Poor
Cores				
Conventional	Saturation measurements from fresh cores	Core must be cut while drilling holes	Yes	Poor
	Lab flooding techniques, imbibition, centrifuge,	Core must be cut while drilling holes	Yes	Poor to Fair
Pressure etc.	Core with specially designed mud	Core must be cut while drilling holes	Yes	Poor to Excellent
Single-well tracer	Backflow hydrolyzed tracers	Yes	Yes	Good to Excellent
Logging tools				
Resistivity	Conventional	No	Yes	Poor
	Log-inject-log	No	Yes	Good to Excellent
Pulsed neutron capture	Conventional	Yes	Yes	Poor
	Log-inject-log with waterflood	Yes	Yes	Good to Excellent
	Log-inject-log with chemicals	Yes	Partially	Fair to Good
Nuclear magnetism[2]	Inject-log	No	Yes	Excellent
Carbon/Oxygen	Conventional	Yes	Yes	Poor
Gamma radiation	Log-inject-log	Yes	No	Unknown (but could be excellent)
Dielectric constant	Conventional	No	Partially	Unknown

1. Expected accuracy at 2 standard deviations-percent pore volume: Excellent: 0-4, Good: 4-8, Fair: 8-12, Poor: greater than 12.
2. Subject to "well-bore" errors, because of shallow penetration.

Each method has advantages and disadvantages, and we would normally use several methods to check and confirm the estimated values. Each method also has its own sampling volume (i.e., near-well to an entire reservoir average). Table 7-5 gives a relative comparison of the expected accuracy for each method as well as its necessary application data.

Please use the expected accuracy given in Table 7-5 with caution, since your actual results will be only as reliable as your data. In addition, since each method samples a different portion of the reservoir, they will each have their own unique uses. Core analysis and logging can provide accurate estimates of residual oil saturation and its vertical distribution, but only in the vicinity of the wellbore, where conditions can be influenced by production or injection well practices. Material balance calculations give one average value of oil saturation for an entire reservoir. Simulation studies improve on this, giving an areal oil saturation distribution and realistically taking past production practices into account. However, the simulator's accuracy depends on an accurate reservoir description (which may be obtained by history matching) and on the accuracy of well production and injection histories. Pressure transient testing also samples a large reservoir area, but requires a knowledge of fluid and rock compressibilities and cannot yield a single, unique interpretation in cases where three fluid saturations (oil, water and gas) are present. Tracer methods can be designed to test large, predetermined volumes of a reservoir. The method gives the averaged oil saturation for the tested zone, and weighs the results to favor saturation remaining in the reservoir's more permeable zones.

As we have already noted, we need to confirm our results using several methods. Since oil saturation can vary widely from one area to another, the residual oil saturation value at a single well, regardless of that value's accuracy, should not be used to decide the fate of a project. In small pilot areas, we want to define the distribution in the entire pattern area, which may involve running several well tests. On a field-wide basis, we will want saturation estimates for the entire field. Although this will have to be done on a less refined scale than a pilot, the sampling method should be detailed enough to provide reasonable field-wide values. Also, extensive pilot work may prove that one or two methods yield reliable data. In that case, the field survey could be done using only these methods.

Water and Gas Saturation

Oil saturation impacts economic success directly, while water and gas saturations influence economics indirectly, and are important primarily from a process mechanism and operational standpoint. Water chemistry dictates whether the injection water is compatible with the existing injection fluid make-up. A small, uniformly distributed gas saturation (below the critical value or even slightly above it) will not adversely affect the performance of a waterflood. However, the presence of a gas cap or a uniformly distributed high gas saturation can control the movement of fluid (injected and in situ) in the reservoir and the "fill-up" volume (i.e., the time required to fill the free gas volume in the reservoir), and result in project failure.

The methods available for determining water and gas saturations are generally the same as those for residual oil saturation, although some for specific methods (i.e., well logging) require modifications. The same advantages and limitations also hold. The tracer method gives only the sum of water and gas saturation, and can be adversely affected by obtaining only oil saturations that are heavily weighted toward the gas-invaded zones.

Fluid Transport Properties and Phase Equilibrium Data

Fluid transport properties (density, viscosity, solubility data, etc.) are used to determine the relative feasibility of a waterflood, and are critical parameters in numerical simulation work conducted for design purposes.

These data are of critical importance in predicting waterflood performance.

Local Rock Properties

A knowledge of local rock properties, such as absolute permeability, porosity, net pay or formation thickness, formation compressibility, and relative permeability and capillary pressure data, is necessary in order to evaluate the feasibility of a recovery process. These data are also useful in the design, implementation and evaluation stages of a project. Local (near-well) permeability controls fluid injectivity/producibility and the product of interwell permeability thickness (kh) affects interwell fluid mobility. These, in turn, determine whether a given process is applicable to a reservoir (i.e., thermal methods have kh/μ requirements, etc.) and directly influence the operating procedures (i.e., injection and production rates, heat losses, etc.) that govern a project's life.

Methods available for determining these rock properties include:

- Core analysis
- Well logging
- Pressure transients, including fall-off, build-up and interference tests

As we saw in the case of residual oil determination, the core and logging methods give detailed vertical distributions of porosity and permeability, but only in the immediate vicinity of the well.

Pressure transient analysis gives a weighted average of the transmissibility (kh/μ), diffusivity (k/$\phi\mu c$), and storage capacity (ϕch) of the zone investigated. This zone is much larger than the volume sampled by coring or logging.

We will need data from all of these sources in order to develop the in-depth reservoir characterization needed for process design and evaluation. Data from each source, supplemented with other reservoir data, must be obtained before geologists can construct detailed reservoir maps correlated between wells. Numerical simulation studies also rely on all of these data sources (particularly pressure transient data) to predict and analyze process performance.

Regional Reservoir Rock Properties

Regional is a term which can be used to describe pilot-sized regions or an entire field. Regional rock properties include such characteristics as:

- Directional permeabilities
- Stratification
- Zone continuity
- Formation dip
- Fractures
- Faults

These reservoir properties, which are beyond our control, can significantly affect the performance of a process. A process that seems technically feasible, based on exhaustive and accurate laboratory tests, may prove a complete failure under reservoir conditions if one or more of these regional characteristics are present. Lack of zone continuity can result in large unswept regions or fluid migration to nonprospective zones. Directional permeability trends and/or stratification can result in large unswept areal or vertical reservoir regions (respectively), resulting in low oil recovery. It is critical to characterize these regional properties in any pilot area and/or field of concern.

Careful geological mapping that incorporates data on local rock properties can help us to define regional property variations. However, we are still forced to interpolate between wells and face the obvious attendant problems. The only positive way to assure zone continuity, determine the presence of faults and/or fractures and evaluate directional rock property trends is to conduct well-planned and analyzed pressure transient testing. The testing methods commonly used are *interference tests or pulse tests.* Pressure transient well data (when analyzed with the appropriate numerical simulator) provide an excellent means of characterizing a test area regionally. We can also analyze past recovery processes and simulate the process in detail in order to supplement this method. To this, we can add even more simulation work, including modeling reservoir conditioning displacements (such as a preflush treatment) or running the simulations associated with residual oil determination using chemical tracer tests. High resolution production seismology provides an additional tool for defining zone continuity and required zone thickness. Regional variations in thickness, the presence or absence of faults, and other structural variations can be mapped to aid in the overall description of a reservoir. In summary, no process should be classified as viable, regardless of successful laboratory prognosis, until the influence of the regional reservoir properties present has been considered.

Regional Fluid Distribution and Dynamic Reservoir Conditions

Regional fluid distribution refers to the presence or absence of water zones or aquifers and gas caps. *Dynamic reservoir conditions* refers to the potential for regional fluid migration in a given test area. Again, the presence or absence of water or gas zones can be determined by the same methods used to determine regional rock properties. Regional fluid migration potential, established from reservoir pressure gradients induced by production/injection practices, could significantly affect fluid movement in the test area and result in nonuniform injection/production ratios for pilot wells and fluid migration out of the test area. Measuring this effect requires us to use numerical simulators and the modeling of past reservoir performance, or data gathered from reservoir pilot floods and/or pressure transient analysis tests.

Process and Operations Design

Process and operations design relies on the integration of detailed laboratory experimental work, reservoir characterization data, field injectivity and pilot tests, and comprehensive waterflood simulation studies. The experimental laboratory work is designed to articulate the process physics, and the simulation work is designed to represent actual reservoir characteristics and their influence on the process.

After assembling the appropriate laboratory data, we need to consider our design from the standpoint of actual field implementation and operations. Here is the point where we incorporate the real field description into the design. Consequently, this work is usually done after some or all of the reservoir characterization work is complete. However, the influence of certain key parameters on process performance could be investigated prior to or during reservoir characterization work, using our best estimate of representative reservoir and fluid properties.

Our concerns at this stage of the design process include:

- pattern type, location and size.
- project duration.
- injected fluid design (i.e., water-injection type, volume, and compatibility) and operations guidelines.
- analysis of reservoir performance and oil recovery prediction.

Each of these factors is important enough to warrant some further exploration here.

Pattern Type, Location and Size

Generally, when we want to evaluate a process that will use a regular well pattern, we would like the test area to have enough wells to completely confine the principal pilot area. Figures 7-1 and 7-2 shows such a well distribution for the five-spot pattern. This is often economically unfeasible, and so isolated patterns are selected (as shown by the patterns in Figure 7-1).

Normal isolated patterns offer the best combination of interpretive advantages and economy of operation. A normal isolated pattern has a weaker dependence on the ratio of fluid injection to production and on mobility than the *inverted isolated pattern* (also shown in Figure 7-2). Every reservoir is unique, however, and other well patterns may be needed. In all cases, the movement of fluid within the pilot area and its imme-

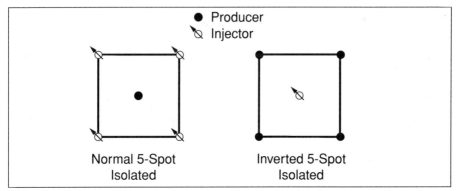

Figure 7-1 Pattern Type, Location, and Size—Isolated

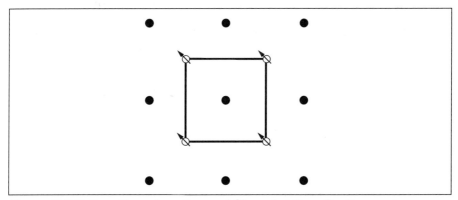

Figure 7-2 Pattern Type, Location and Size—1st Row Confinement

diate surroundings should be simulated, using actual reservoir conditions, to see whether or not injected fluid is confined to the test area.

Site locations for a pilot test should be governed by the main purpose of every pilot, which is (as we have seen) to maximize our chances of getting a meaningful interpretation. We should try to avoid test locations which may pose pattern fluid containment problems. These problems could result from a lack of zone continuity or from the strong influence of surrounding producers. We should try to choose a homogeneous, representative reservoir section in order to avoid interpretation problems and to provide some basis for extrapolating results to a field-wide scale. This selection process could disclose more than one suitable pilot site. These sites should then be evaluated using preliminary tests, such as pressure transient testing, before making a final selection.

Selection of a pattern size is complicated by the influence of a number of conflicting requirements. First, to minimize pilot testing costs, we would like the pilot area to be as small as possible without losing its ability to represent the reservoir as a whole. On the other hand, preliminary economic analysis may dictate a minimum pattern size to ensure that subsequent full-field implementation is profitable. If the process's performance viability is influenced by pattern size (as in a steam drive, where heat loss may be an important factor, or in miscible or surfactant flood processes, where we need to ensure slug integrity), we are faced with a dilemma. From a practical standpoint, we must use the minimum size dictated by economics, despite the possibility of its incurring higher costs and a longer lead time to implementation. The only alternative would be to place our trust completely in an enhanced recovery simulator. If we are fully confident in our ability to interpret the performance of a small pilot, we can confidently predict the performance of a larger pilot or pattern area that would correspond to the size dictated by economics.

Project Duration

Project duration is clearly tied to the injectivity and productivity of the pattern wells and to the pattern size. We need to run the appropriate pressure buildup and fall-off tests in order to obtain reliable estimates of well injectivity and productivity. These tests would, of course, be conducted during reservoir characterization. We can then use simulation runs to predict project life.

Injected Fluid Design and Operation Guidelines

Injected fluid design refers to the water-injected type, volume and compatibility. *Operation guidelines* refers to the injection/production ratios for the wells in the pattern, completion intervals, well back pressure, duration of water injection periods and specified pressures.

A simulation study is required to provide individual rate schedules for each well. This allows us to insure optimum confinement of the injected fluids to the pilot area. Transmissibility irregularities within a pattern could, for instance, require reduced injection rates in a particular pattern quadrant. Also, regional fluid migration potential caused by offset wells or pattern geometry irregularities will necessitate unbalanced injection/production ratios. The influence of practices such as specialized well completion (i.e., restricted fluid entry) and the effect of well back pressure can be studied and used to develop operating guidelines for field implementation.

Analysis of Reservoir Performance and Oil Recovery Predictions

A simulator is also a reliable tool for predicting process performance. When the injected fluid and operation design studies are complete, we will be left with overall performance and oil recovery data. Simulators developed to design and evaluate waterfloods are readily available in the industry today. Table 2-3 shows some quick evaluation "rules of thumb" for waterflooding. Even if detailed simulations are performed, it is recommended to check them in light of these rules of thumb.

Equipment Design

The special equipment requirements for specific processes have already been mentioned. Our intent here is to examine the need to integrate equipment design work into the overall project design and pilot goals.

Engineers involved with equipment design should participate in all process design and process monitoring discussions. Since the goal of a pilot is to obtain interpretable data, the engineers involved in monitoring and evaluating the process may have special needs. These needs must be passed on to the equipment design group. One example would be the need to measure water production as soon as water breaks through.

Other examples would include the measurement of injection and production well data at bottomhole conditions, which requires special equipment and well completions and must be integrated with the overall equipment design work, as well as:

- equipment to monitor produced fluid rates
- general data acquisition equipment, along with its associated transmission and storage requirements

Pilot tests are not only for testing and validating process concepts; they also offer a unique opportunity to test different equipment options, and develop the maintenance records and operating cost data instrumental to an economic evaluation. Comparison tests can be made on fluid-handling equipment, various well completion procedures (including the use of sand screens, special alloys to withstand corrosion, tubular insulation, etc.) and monitoring and automated control systems. A systematic evaluation procedure then provides the data necessary for selecting equipment and developing maintenance procedures for full-scale projects.

Data Acquisition Design

Effective data acquisition requires the staff responsible for data acquisition to be involved with the overall project design. They must get input from both process and equipment engineers. They should also consider using automated data acquisition facilities, as well as a computerized means of processing and storing data. A similar system should be developed and used to process and store equipment-related data in order to obtain accurate maintenance and operating cost data.

Well data is, of course, useful for regularly monitoring the project, as well as for performing its final engineering analysis. A comparable suite of data for equipment maintenance and operating costs, plus other project costs, will provide the basis for a sound economic evaluation of the project.

Economic Evaluation, Including Project Scale-up

An economic evaluation performed while a project is at the preliminary stage provides the initial justification for detailed project design. Completion of a detailed design, including reservoir characterization, provides an opportunity to do an economic update using more precise performance prediction data and more realistic cost data. If, after this new economic evaluation, the project is still economically viable, the final test or project can proceed.

When process, equipment and data acquisition design have been integrated, smooth project monitoring should result. Data are necessary for both process and economic evaluations. A complete simulation of the process operation, as recommended in the process design stage, provides project operating guidelines and performance data to compare to actual pilot or project performance. Process reevaluation contingency plans should be developed to provide a means of modifying performance using appropriate simulation studies if actual performance deviates from that expected. This ability to revise the process design when necessary thus provides new operating guidelines that could save a project from failure.

Upon completing a project, the final evaluation should provide a basis for confidently predicting the performance of the process under different field conditions. If this is possible, the pilot can be classified as a success. Because of process mechanism complexities and the need to include actual reservoir conditions, this final process evaluation will rely on the simulator used for process design.

A successful pilot should provide all the information necessary for a field-wide implementation of the process. Validating our simulation of the process by using a comprehensive numerical simulator provides a means of implementing the process under field conditions different from those in pilot area. It can thus provide reliable recovery predictions for final economic evaluation of a commercial project, as well as revised operational guidelines, if necessary. A validated simulator also provides the means to test modifications to the process that could improve performance.

Reliable equipment, maintenance and cost data, as well as records of comparative tests on competing equipment, provide a sound basis for selecting the best equipment for a commercial pilot and for carrying out the final economic analysis. The knowledge gained by testing various reservoir characterizations during the pilot project will be used to develop a simple but reliable approach to defining the critical reservoir properties needed for a successful field-wide implementation.

In summary, if the pilot has validated our process interpretation procedure, we can expand the process, confident of its economic success.

Why Waterfloods Fail

Sometimes the actual field performance does not match the predicted values because of inevitable assumptions and rough estimation of some parameters that are not available. Depending upon the estimates of these parameters, the forecasted performance may be much different, and in some cases, may lead to project failures. Therefore, it is a good idea to identify key parameters that have significant effect on the waterflood performance, and determine the sensitivity of these parameters on the results. This process assists the asset management team in evaluating the uncertainties, and helps them focus on carefully collecting accurate information about the pertinent parameters.

The above process is a means of evaluating uncertainties, but it does not guarantee a successful waterflood. Although the design and operation of waterfloods have become well known in the industry, still there are major economic differences as a result of average and exceptional waterfloods. In addition, sometimes even today, some waterfloods fail. Some of the reasons for failure of waterfloods are described in Table 7-6.[7]

Table 7-6 Why Waterfloods Fail (From Jackson, R. "Why Waterfloods Fail," World Oil, March, 1968, p 65)

- Poor Sweep Efficiency (45%)
  - Vertical permeability variations causing early breakthrough and high water production
  - Fractures and directional permeability
  - Fluid distribution in thick reservoirs with high vertical permeability; water underruns the oil
  - Viscous fingering and poor mobility ratio
  - Unbalanced injectivity
- Unexpected Expenses (45%)
  - Extensive remedial work
  - Equipment failure and underdesign of producing and injection equipment
- Others (10%)
  - Initial oil saturations too small to form an oil bank
  - Oil resaturation of gas cap

An Example Of Waterflood Design

A field that was discovered many years ago is now depleted. It consists of a simple domal structure, and five of the nine wells drilled were producers (Figure 7-3). Primary producing mechanisms were fluid and rock expansion, (reservoir pressure above the bubble point), solution gas drive and limited natural water drive. Data available are limited; even the gas, oil and water production data are unreliable. Reservoir pressures were not monitored.

An integrated team of geoscientists and engineers was charged by the management to review the past performance and investigate waterflood potential of this field which is akin to real life reservoirs. The team's approach was to

- build an integrated geoscience and engineering model of the reservoir using available data and correlations
- simulate full-field primary production performance without history matching since no historical pressure data are available
- forecast performance under peripheral and pattern waterflood

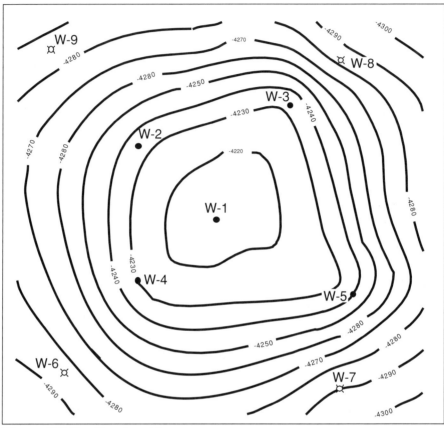

Figure 7-3 Top Structure Map

This chapter presents the results of the study. Even though the field is not real, much can be learned about how to engineer a waterflood project even with incomplete data.

Reservoir Data

An analysis of the logs from the nine wells showed that the reservoir heterogeneity can be represented by five producing horizons. Permeability was computed from a correlation of porosity vs. permeability. Permeability in the x and y directions are considered to be the same, i.e., no directional permeability. The vertical to horizontal permeability ratio is assumed to be 0.1. Layer properties are presented in the figures as listed next:

Tables 7-7 through 7-11   Structure tops, gross and net thickness, porosity and permeabilities for layers 1-5

Figures 7-4 through 7-7   Gross and net thickness, porosity, and permeability of layer 1 as an example

Figure 7-8   Cross-section

The reservoir contains 33° API crude oil at 2,332 psia original reservoir pressure and 123°F temperature. It is undersaturated, having the initial pressure several hundred psi above the bubble point (1,855 psia). Fluid properties and gas-oil and water-oil relative permeability data, were obtained from correlations (See Figures 5-2 through 5-5).

Table 7-7 Reservoir Description of Layer 1

Well No.	HTOP	GROSS (ft)	NET (ft)	PHI (%)	Kx (md)
1	-216	12.0	12.0	24.0	64
2	-235	10.5	10.5	21.0	50
3	-233	9.5	9.5	19.0	42
4	-230	11.0	11.0	22.0	54
5	-240	9.0	9.0	18.0	39
6	-285	7.5	7.5	15.0	30
7	-290	6.5	6.5	13.0	26
8	-283	7.0	7.0	14.0	28
9	-286	8.0	8.0	16.0	33

Table 7-8 Reservoir Description of Layer 2

Well No.	HTOP	GROSS (ft)	NET (ft)	PHI (%)	Kx (md)
1	n/a	16.0	16.0	23.0	59
2	n/a	14.0	14.0	20.0	46
3	n/a	12.5	12.5	18.0	39
4	n/a	14.5	14.5	21.0	50
5	n/a	12.0	12.0	14.0	36
6	n/a	10.0	10.0	14.0	28
7	n/a	8.5	8.5	12.0	24
8	n/a	9.5	9.5	13.0	26
9	n/a	10.5	10.5	15.0	30

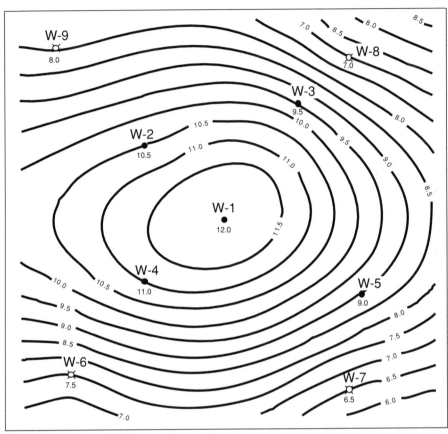

Figure 7-4 Gross Thickness-Layer 1

Table 7-9 Reservoir Description of Layer 3

Well No.	HTOP	GROSS (ft)	NET (ft)	PHI (%)	Kx (md)
1	n/a	21.0	21.0	22.0	54
2	n/a	18.5	18.5	19.0	42
3	n/a	16.5	16.5	17.0	36
4	n/a	19.5	19.5	20.0	46
5	n/a	15.5	15.5	16.0	33
6	n/a	13.0	13.0	14.0	28
7	n/a	11.5	11.5	12.0	24
8	n/a	12.5	12.5	13.0	26
9	n/a	14.0	14.0	15.0	30

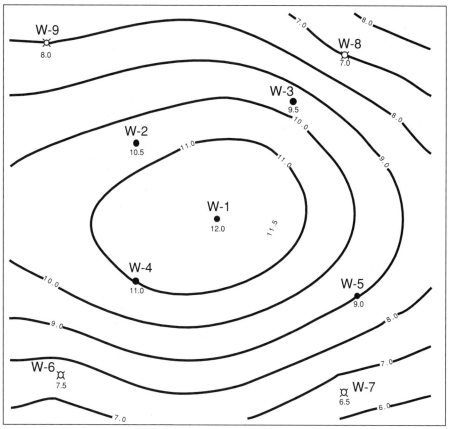

Figure 7-5 Net Thickness-Layer 1

Table 7-10 Reservoir Description of Layer 4

Well No.	HTOP	GROSS (ft)	NET (ft)	PHI (%)	Kx (md)
1	n/a	11.0	11.0	22.0	50
2	n/a	9.5	9.5	19.0	39
3	n/a	8.5	8.5	18.0	36
4	n/a	10.0	10.0	20.0	42
5	n/a	8.5	8.5	16.0	33
6	n/a	7.0	7.0	14.0	26
7	n/a	6.0	6.0	12.0	22
8	n/a	6.5	6.5	13.0	24
9	n/a	7.5	7.5	15.0	28

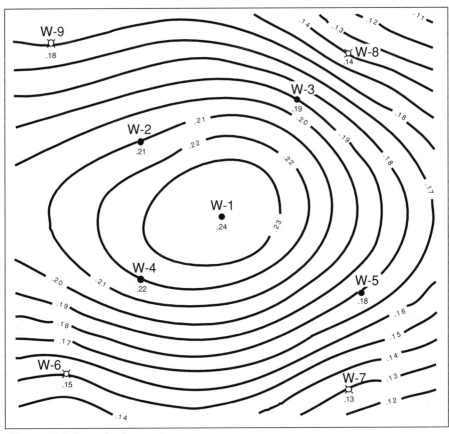

Figure 7-6 Porosity-Layer 1

Table 7-11 Reservoir Description of Layer 5

Well No.	HTOP	GROSS (ft)	NET (ft)	PHI (%)	Kx (md)
1	n/a	13.0	13.0	20.0	46
2	n/a	11.5	11.5	17.0	36
3	n/a	10.5	10.5	16.0	33
4	n/a	12.0	12.0	18.0	39
5	n/a	10.0	10.0	15.0	30
6	n/a	8.0	8.0	12.0	24
7	n/a	7.0	7.0	11.0	22
8	n/a	7.5	7.5	12.0	26
9	n/a	8.5	8.5	13.0	24

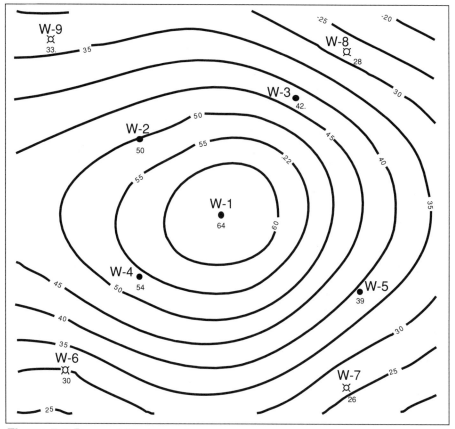

Figure 7-7 Permeability-Layer 1

Reservoir Modeling

Using a black oil simulator with 25x25x5 grids (3,125 cells, with 1.02 acres each) a full-field reservoir simulation model was constructed to predict primary performance. The model used the following well production limitations:

- Economic oil production rate, STB/day   10
- Maximum gas-oil ratio, SCF/STB   2,500
- Maximum water cut, %   95
- Minimum bottom hole pressure, psia   150
- Completed layers   1, 2, 3

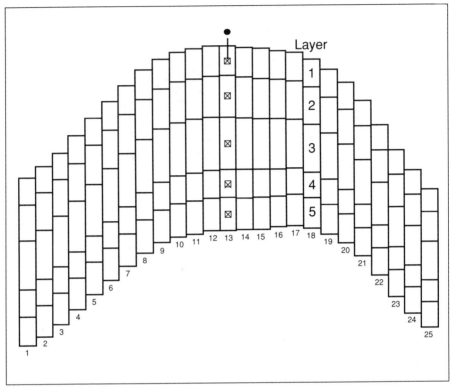

Figure 7-8 Cross Section

The original oil-, gas- and water-in-place were computed to be 26.2 MMSTB, 10.8 BSCF and 19.1 MMSTB. The reservoir pore volume was 50.1 MMRB. Primary oil recovery was 4.1 MMSTB (15.7 % OOIP) after 8.3 years. Cumulative oil productions from the individual wells were:

Well	Cumulative Oil Production	Status
1	1,418.0 MSTBO	
2	604.9 MSTBO	Shut-in due to high GOR
3	687.1 MSTBO	Shut-in due to high GOR
4	930.9 MSTBO	
5	467.2 MSTBO	Shut-in due to high GOR

Figures 7-9 and 7-10 show field-wide production performance.

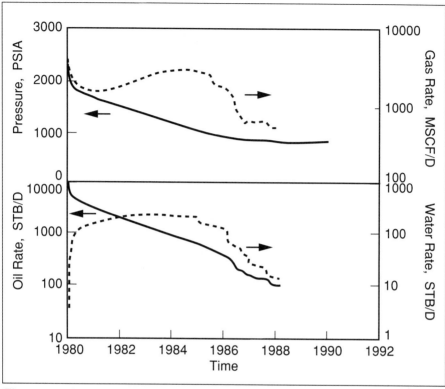

Figure 7-9 Fieldwide Primary Production Rates and Pressure Behavior

Production Rates and Reserves Forecasts

Performance forecasts for waterflood recovery were made for the following operating scenarios in order to determine the optimum development plan:

1. Peripheral waterflood, using the existing 5 producing wells and the 4 dry holes at the periphery as injectors (Figure 7-11)
2. Enhanced peripheral waterflood, with 8 injectors and 9 producers (Figure 7-12)
3. Single 5-spot pattern waterflood, using the central existing producing well and converting the 4 surrounding wells to injectors (Figure 7-13)

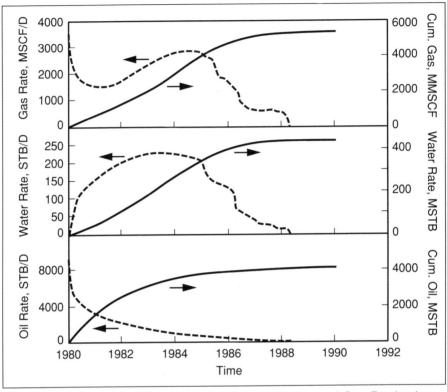

Figure 7-10 Fieldwide Primary Cumulative Oil, Water, and Gas Productions

4. Four 5-spot pattern waterfloods, using 9 injectors and 4 producers (Figure 7-14)
5. Multiple rows of direct line drive waterfloods, using 13 injectors and 12 producers (Figure 7-15)

Calculated results for these cases are presented as listed below:

Table 7-12   Compares 10 years of annual waterflood oil productions

Figure 7-16   Compares 10 years of primary and 20 years of waterflood oil productions

Figure 7-17   Compares waterflood oil recovery vs. pore volume of water injected

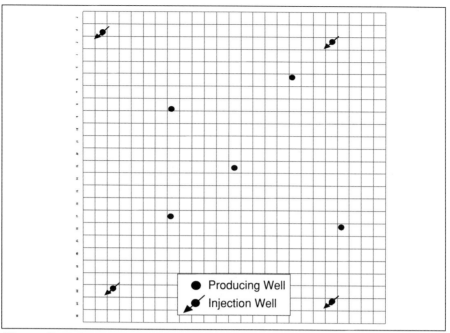

Figure 7-11 Case 1 Development Plan

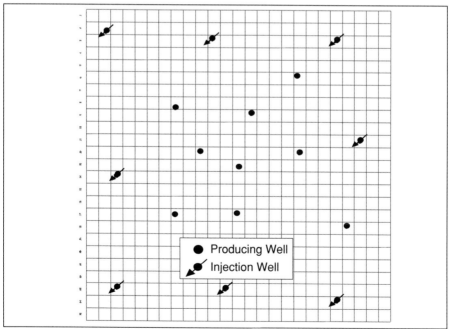

Figure 7-12 Case 2 Development Plan

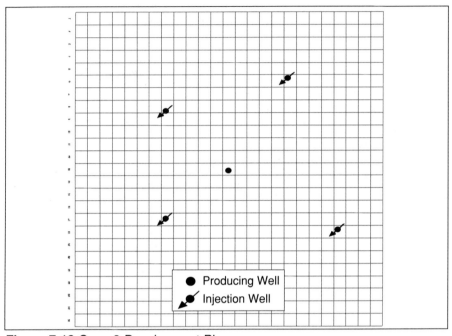

Figure 7-13 Case 3 Development Plan

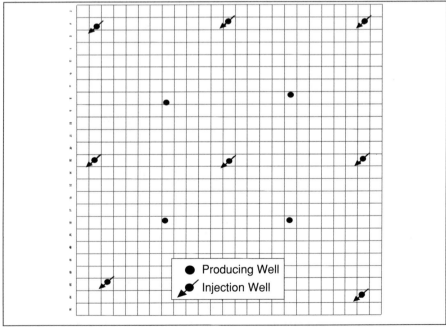

Figure 7-14 Case 4 Development Plan

Table 7-12 Produced Oil Volumes

Year	Case 1	Case 2	Case 3	Case 4	Case 5
1	69.8	275.2	37.5	82.0	279.8
2	49.8	189.7	27.1	127.4	786.2
3	48.7	451.5	27.5	321.0	882.4
4	51.6	656.3	28.3	325.8	605.9
5	55.0	605.7	30.0	293.1	486.8
6	60.8	515.6	35.7	266.5	405.6
7	134.8	460.6	54.0	248.8	341.0
8	234.7	407.6	93.5	237.0	288.9
9	242.1	352.0	155.5	228.1	253.7
10	241.4	305.1	221.1	221.3	220.3
10 Year Total	996.0	4,219.5	710.2	2,351.3	4,550.6

Project life being the same, Table 7-12 and Figures 7-16 and 7-17 show significantly higher recoveries in Cases 2 and 5, which have more injectors and producers than the other cases. Case 3, with only one producer shows the least recovery. Or in other words, recoveries are directly related to the number of production wells.

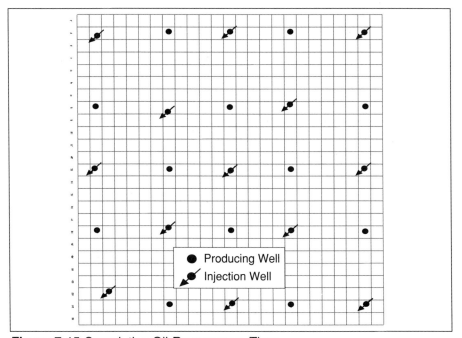

Figure 7-15 Cumulative Oil Recovery vs. Time

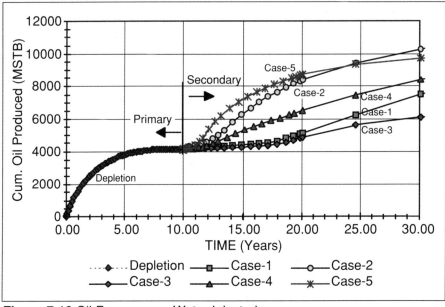

Figure 7-16 Oil Recovery vs. Water Injected

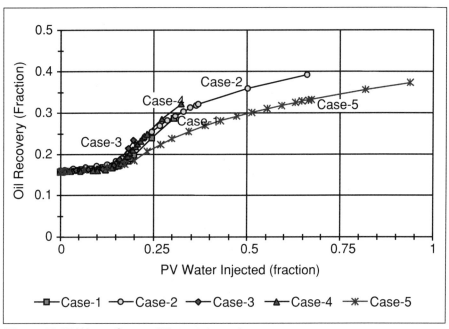

Figure 7-17 Water Cut vs. Water Injected

More oil recovery with more injection and production wells may not provide the best economically viable scheme. That can only be determined by economic analysis of the cases considered in chapter 11. Also, the limited cases considered here do not necessarily give the answer as to how best to develop the field for waterflooding. These examples simply demonstrate that more than one way of developing the field needs to be investigated in order to determine the potentially most economically viable project.

References

1. Thakur, G. C. "Reservoir Management of Mature Fields," IHRDC Video Library for Exploration and Production Specialists (1992).

2. Willhite, G. P. "Waterflooding," SPE Textbook Series, Volume 3 (1986).

3. Craig, F. F., Jr. "The Reservoir Engineering Aspects of Waterflooding," SPE Monograph Volume 3, Richardson, TX (1971).

4. Rose, S. C., J. F. Buckwalter, and R. J. Woodhall. "The Design Engineering Aspects of Waterflooding," Monograph Series, SPE, Richardson, TX (1989): 11.

5. Thakur, G. C. "Waterflood Surveillance Techniques – A Reservoir Management Approach," *JPT* (October 1991): 1180-88.

6. Satter, A. and G. C. Thakur. "Integrated Petroleum Reservoir Management: A Team Approach," *PennWell Books*, Tulsa, OK (1984).

7. Jackson, R. "Why Waterfloods Fail," World Oil (March 1968): 65.

8

Waterflood Production Performance and Reserves Forecast

Waterflooding is usually initiated in pressure depleted or nearly depleted reservoirs. Most commonly, solution gas drive reservoirs develop free gas saturation due to pressure depletion. Initially, the reservoir pressure is restored as the gas-filled pore volumes are refilled with the injected water, redissolving free gas back into the oil. The oil production response occurs after the fill-up of the gas space. The injected water eventually breaks through at the producing wells; generally, very little water is produced before the peak oil production rate is reached if the reservoir resembles a homogeneous formation. The timing of the oil response and water breakthrough, and the magnitude of the peak oil production rate depend upon the reservoir characteristics and the injection rate. After the peak production rate is reached, oil production rate declines with increase in water cut. Figure 8-1 illustrates the performance curve of a typical successful waterflood.[1]

A major waterflood asset management activity is to estimate reserves, recovery rates, and flood life for designing a project, and also to analyze past and future performance during the life of the flood for monitoring and evaluation purposes.

Commonly used techniques are:

- Volumetric method
- Empirical method
- Classical method
- Performance curve analyses
- Simulation method

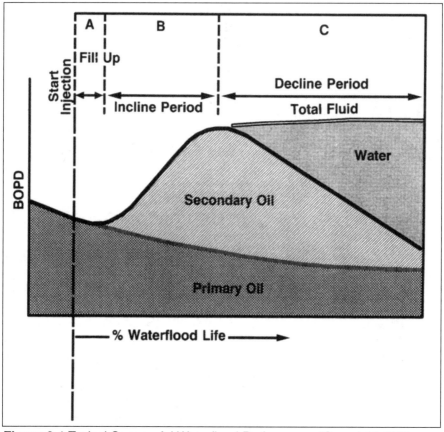

Figure 8-1 Typical Successful Waterflood Performance (Copyright 1991, SPE, from JPT, October 1991)

Even though reservoir engineers are mainly involved in this activity, geologists, petrophysicists and production engineers are also responsible for providing data, reviewing and verifying results of the reservoir performance analysis.

This chapter presents techniques used for waterflood performance analysis and reserves forecasts.

Volumetric Method

Ultimate waterflood oil recovery = Oil-in-place prior to waterflood x recovery efficiency

Oil-in-place prior to waterflood can be estimated by subtracting the cumulative production from the original-oil-in-place which is given by the bulk volume of the reservoir, the porosity, the initial oil saturation and the oil formation volume factor as shown below:

$$N = \frac{7758 \, Ah\phi S_{oi}}{B_{oi}} \qquad (8\text{-}1)$$

where:

$$7758 = \frac{43,560 \, ft^2 / Ac}{5.614 \, ft^3 / bbl}$$

N	= OOIP, STB
A	= area, acre
h	= average thickness, ft (oil interval)
ϕ	= average porosity, fraction
S_{oi}	= initial oil saturation, fraction
B_{oi}	= initial oil formation volume factor, RB/STB
RB	= reservoir barrel
STB	= stock tank barrel

The bulk volume is determined from the isopach map of the reservoir, average porosity and oil saturation values from log and core analysis data and oil formation volume factor from laboratory tests or correlations.

If the waterflood is initiated while the reservoir is still in primary production mode, adjustment to ultimate waterflood recovery has to be made to account for the remaining primary production.

Once the original oil in place has been estimated, the ultimate recovery can be estimated using recovery efficiency factor. The oil recovery efficiency factor (STB/acre-ft or % OOIP) may be estimated from the performance data on similar and/or offset reservoirs. It can be estimated also by multiplying the displacement efficiency, determined from the laboratory flood pot tests, by an estimated volumetric sweep efficiency. Laboratory core floods, ideally using representative formation cores and actual reservoir fluids, are the preferred method for obtaining of waterflood residual oil saturation and displacement efficiency.

Fractional flow theory (see Appendix D) can be used to estimate residual oil saturation and displacement efficiency, but it requires measured water-oil relative permeability curves.[2] Alternately, empirical correlations such as Croes and Schwarz based upon the results of laboratory waterfloods can be also used to calculate the displacement efficiency (Figure 8-2). [3]

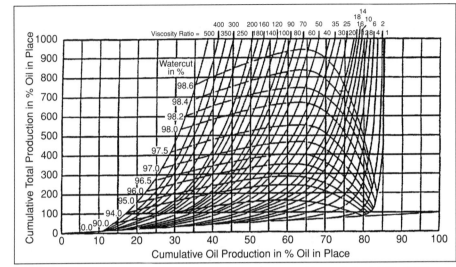

Figure 8-2 Experimental Waterflood Performance (Copyright 1955, SPE, from Trans. AIME, 1955)

Although the volumetric method gives an estimate of the expected waterflood recovery, it does not provide oil production rate vs. time performance which is needed for economic evaluation of a waterflood project. However, volumetric method is useful for initial project screening for capital investment possibilities, for evaluating potential waterflood value, and for prioritizing waterflood prospects by expected ultimate recovery.

Empirical Methods

Empirical methods for predicting waterflood recovery are based upon:

1. Correlations with rock and fluid properties
2. Rules of thumb for obtaining peak oil production rate, time to first oil production response, the oil production rate decline from the peak rate, etc

Examples of empirical approaches are the methods proposed by Guthrie-Greenberger, Arps, et, al., Shauer, Guerrero and Earlougher, and Bush and Helander.[4,5,6,7,8] Craig provided a summary of these methods which can give reasonable results when developed from local field performance and applied in the same general area.[9] The methods with a provision for time in the predicted performance by assuming injection rate can be useful for estimating project life, designing and planning, and economic analyses.

Classical Methods

Craig summarizes many published classical methods for predicting waterflood performance.[9] Commonly used prediction methods, primarily concerned with reservoir heterogeneity but considering piston-like displacement, are

1. Dykstra-Parsons Method based upon a correlation between waterflood recovery and both mobility ratio and permeability variation factor (Figure 8-3)[10]
2. Stiles Method accounting for the different flood-front positions in liquid-filled, linear insulated layers having different permeabilities.[11] Permeability variation of the layers and the layer flow capacities are used to derive oil recovery and water cut equations
3. Prats-Matthews-Jewett-Baker Method based upon a correlation of oil recovery, including the combined effects of mobility ratio and areal sweep efficiency, and considering the presence of free gas prior to waterflooding and variation in injectivity through the life a flood[12]

Commonly used prediction methods primarily concerned with displacement mechanisms are:

1. Buckley-Leverett Method considering the immiscible displacement of oil by water in a linear or a radial system.[2] Welge modification to the frontal advance equation greatly simplified its use[13]
2. Craig-Geffen-Morse Method based upon a modified Welge equation and correlations of areal sweep efficiency at and after breakthrough[14]
3. Higgins and Leighton Method based upon stream tube approach at unit mobility ratio, shape factors (relative length and cross-sectional area for equal volume segments), and Buckley-Leverett displacement mechanisms[15]

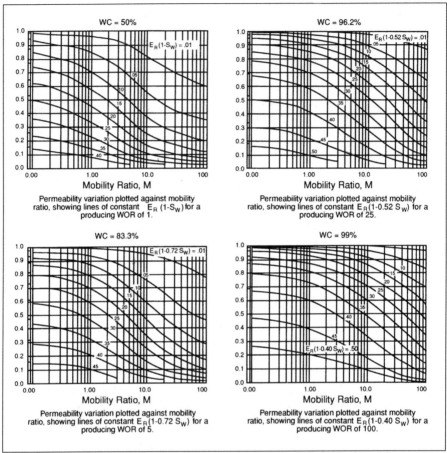

Figure 8-3 Dykstra-Parsons Waterflood Recovery Correlations (After Secondary Recovery of Oil in the United States, second edition, API, 1950[10])

The Craig, et al. method is one of the most thorough and practical prediction methods available for 5-spot pattern. Higgins-Leighton method is applicable for the 5-spot, 7-spot, direct and staggered line drive pattern, and peripheral patterns.

Table 8-1 lists the features of Dykstra-Parsons, Stiles, Prats, et al., Buckley-Leverett and Craig, et al. prediction methods. It can be seen that these methods which are based upon many restrictive assumptions are far from being realistic. Figure 8-4 shows an example of predicted performances of several empirical and classical methods compared to the actual results from a waterflood.[7] It shows the wide divergence in accuracy and prediction.

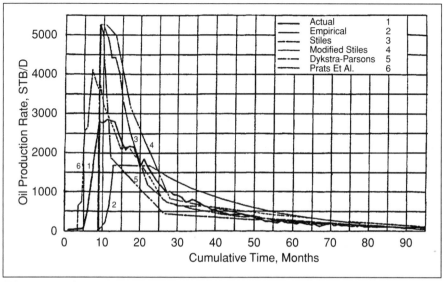

Figure 8-4 Comparison of Actual and Predicted Waterflood Recovery

Table 8-1 Features of Classical Waterflood Prediction Methods (Abdus Satter and Ganesh Thakur, Integrated Petroleum Reservoir Management: A Team Approach, Copyright PennWell Books, 1994)

	Dykstra-Parsons	Stiles	Prats et al.	Buckley-Leverett	Craig et al.
Linear Flow	x	x	x		
Piston-Like Frontal				x	x
Cross Flow					
Areal	no	no	no	no	no
Vertical	no	no	no	no	no
Initial Gas Saturation	no	no	yes	no	yes
Sweep					
Areal	no	no	yes	no	yes
Vertical	yes	yes	yes	no	yes
Mobility Ratio	yes	1.0	yes	yes	yes
Stratification	yes	yes	yes	no	yes
Pattern					
5-Spot	no	no	yes	no	yes
Other	no	no	no	no	no

Performance Curve Analyses

When sufficient production data are available and production is declining, the past production curves of individual wells, lease or field can be extended to indicate future performance. The very important assumption in using decline curves is that all factors that influenced the curve in the past remain effective throughout the producing life. In reality, many factors influence production rates and consequently decline curves; these are proration, changes in production methods, workovers, well treatments, pipeline disruptions, weather and market conditions. Therefore, care must be taken in extrapolating the production curves in the future. When the shape of a decline curve changes, the cause should be determined and its effect upon the reserves evaluated.

The commonly used performance curve analysis methods for waterflood projects are:

1. Log of production rate vs. time (Figure 8-5)
2. Production rate vs. cumulative production (Figure 8-6)
3. Log of water cut or oil cut vs. cumulative production
    (Figures 8-7 and 8-8)

When Type 1 and 2 plots are straight lines, they are called constant rate or exponential decline curves. Since a straight line can be easily extrapolated, exponential decline curves are most commonly used. In case of harmonic or hyperbolic rate decline, the plots show curvature.

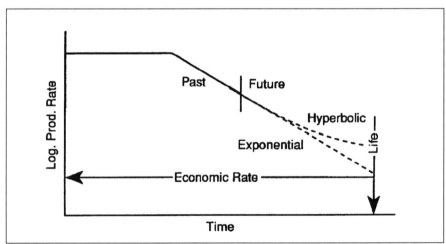

Figure 8-5 Log of Production Rate vs. Time (Abdus Satter and Ganesh Thakur, Integrated Petroleum Reservoir Management: A Team Approach, Copyright PennWell Books, 1994)

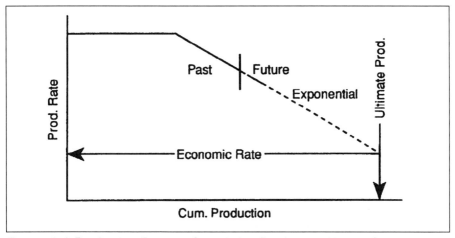

Figure 8-6 Production Rate vs. Cumulative Production (Abdus Satter and Ganesh Thakur, Integrated Petroleum Reservoir Management: A Team Approach, Copyright PennWell Books, 1994)

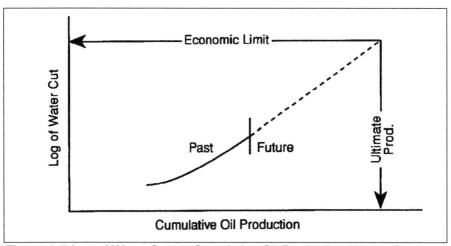

Figure 8-7 Log of Water Cut vs. Cumulative Oil Production (Abdus Satter and Ganesh Thakur, Integrated Petroleum Reservoir Management: A Team Approach, Copyright PennWell Books, 1994)

Both the exponential and harmonic decline curves are the special cases of the hyperbolic decline curves (Figure 8-5). Unrestricted early production from a well shows hyperbolic decline rate. However, constant or exponential decline rate may be reached at a latter stage of production.

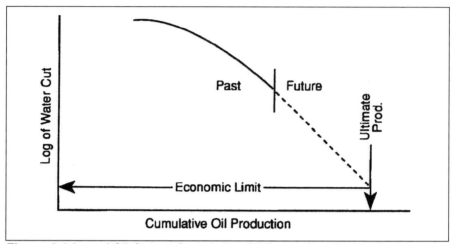

Figure 8-8 Log of Oil Cut vs. Cumulative Oil Production (Abdus Satter and Ganesh Thakur, Integrated Petroleum Reservoir Management: A Team Approach, Copyright PennWell Books, 1994)

Mathematical expressions for production rates, and cumulative productions for the hyperbolic, harmonic and exponential declines are presented by Arps.[16,17]

A general mathematical expression for the rate of decline, D, can be expressed as

$$D = -\frac{dq/dt}{q} = Kq^n \qquad (8\text{-}2)$$

where:

q = production rate, barrels per day, month or year
t = time, day, month or year
K = constant
n = exponent

The decline rate in Equation (8-2) can be constant or variable with time yielding three basic types of production decline as follows:

1. Exponential or constant decline

$$D = -\frac{dq/dt}{q} = K = -\frac{ln\left(\frac{q_i}{q_i}\right)}{t} \qquad (8\text{-}3)$$

when:

$n \quad = \quad 0, K = \text{constant}$

$q_i \quad = \quad$ initial production rate

$q_t \quad = \quad$ production rate at time t

The rate-time and rate-cumulative relationships are given by

$$q_t = q_i \cdot e^{-Dt} \qquad (8\text{-}4)$$

$$Q_t = \frac{q_i - q_t}{D} \qquad (8\text{-}5)$$

where:

$Q_t \quad = \quad$ cumulative production at time t

A familiar rate constant for exponential decline is as follows:

$$D' = \frac{\Delta q}{q_i} \qquad (8\text{-}6)$$

where Δq is the rate change in the first year. In this case, the relationship between D and D' is given below:

$$D = - \ln\left(1 - \frac{\Delta q}{q_i}\right) = - \ln(1 - D') \qquad (8\text{-}7)$$

2. Hyperbolic decline

$$D = - \frac{dq/dt}{q} = Kq^n \ (0<n<1) \qquad (8\text{-}8)$$

Note that this is the same equation as the general decline rate equation (Equation 8-2) except for the constraint on n.
For initial condition

$$K = \frac{D_i}{q_i^n}$$

The rate-time and rate-cumulative relationships are given by:

$$q_t = q_i(1+ nD_i t)^{-\frac{1}{n}} \qquad (8\text{-}9)$$

$$Q_t = \frac{q_i^n(q_i^{1-n} - q_t^{1-n})}{(1 - n)D_i} \qquad (8\text{-}10)$$

where D_i = initial decline rate

3. Harmonic decline

$$D = - \frac{dq/dt}{q} = Kq \qquad (8\text{-}11)$$

when $n = 1$

For initial condition

$$K = \frac{D_i}{q_i}$$

The rate-time and rate-cumulative relationships are given by:

$$q_t = \frac{q_i}{(1 + D_i t)} \qquad (8\text{-}12)$$

$$Q_t = \frac{q_i}{D_i} \ln \frac{q_i}{q_t} \qquad (8\text{-}13)$$

Type 3 curves are employed when economic production rate is dictated by the cost of water disposal. A straight line extrapolation of log of water cut versus cumulative oil production may not be reasonably done in the higher water cut levels. It may yield a conservative estimate of reserves. On the other hand, if oil cut data are used instead of water cut in the same levels, straight line extrapolation of log of oil cut versus cumulative oil production may deteriorate and lead to optimistic reserve estimates.

An example is presented to illustrate application of decline curve analysis. Figure 8-9 shows oil, water, and gas production performance of a well. Figure 8-10 shows past oil production history match (log of oil production rate vs. time) and prediction of reserves. An analysis gave exponential decline at 21.33 % per year. Based on an economic production rate of 10 STBO per day, the reserves were calculated to be 32.3 thousand barrels of oil at the end of 2002.[5] Figure 8-11 shows a plot of water cut vs. cumulative oil production. It is emphasized that both oil rate and water cut should be analyzed to ensure reliability in the results.

Because extrapolation of the past water cut plot is often complicated, Ershaghi, et al[18] devised a method to plot recovery efficiency vs. X, as defined below, which yielded a straight line:

$$E_R = mX + n \qquad (8\text{-}14)$$

where

E_R = over-all recovery efficiency

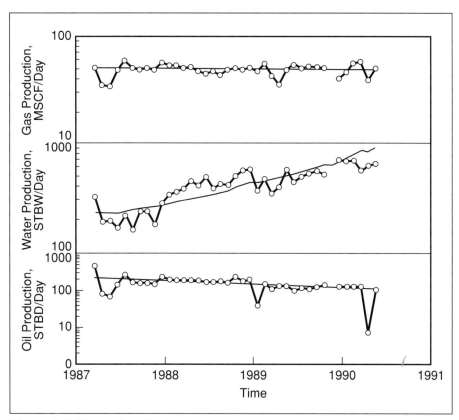

Figure 8-9 Production Performance

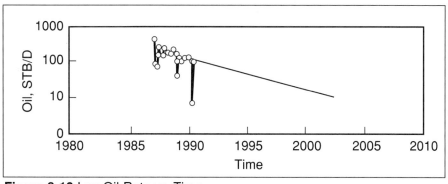

Figure 8-10 Log Oil Rate vs. Time

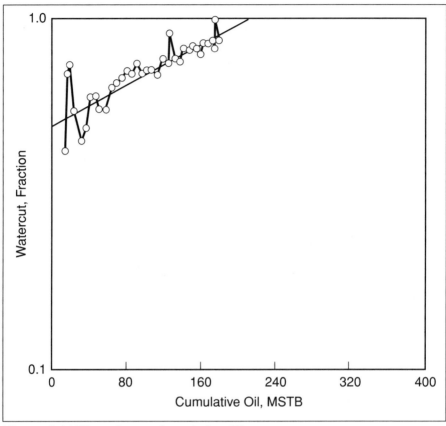

Figure 8-11 Water Cut vs. Cumulative Oil

$$X = - \, [ln(1/f_w - 1) \, -1/f_w] \qquad\qquad (8\text{-}15)$$

This method is more general than the conventional plot of water cut vs. cumulative oil production, and more applicable when water cut exceeds 0.5. Figures 8-12 and 8-13 present an example of the authors' application of their techniques.[18]

Given actual water cut vs. recovery efficiency data, a graph of recovery vs. X would result in a straight line which may be extrapolated to any desired water cut to obtain corresponding recovery. The parameters m and n in Equation (8-14) can be derived from the straight-line relationship in Figure 8-12. These values then can be used in Equation (8-14) to predict water cut vs. recovery in Figure 8-13. It can be clearly seen in Figure 8-13 that extrapolation would result in pessimistic ultimate recovery.

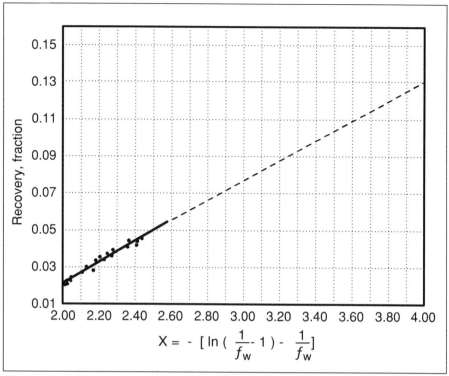

$$X = -\left[\ln\left(\frac{1}{f_w} - 1\right) - \frac{1}{f_w}\right]$$

Figure 8-12 Recovery vs. X

Reservoir Simulation

Numerical reservoir simulators play a very important role in the modern reservoir management process. They are used to develop reservoir management plan, monitor and evaluate reservoir performance, and ultimate recovery of hydrocarbons.

Numerical simulation is still based upon material balance principles, taking into account reservoir heterogeneity and direction of fluid flow. Unlike the classical material balance approach, a reservoir simulator takes into account the locations of the production and injection wells and their operating conditions. The wells can be turned on or off at desired times with specified downhole completions. The well rates or the bottom hole pressures, or even both the rates and pressures can be set as desired.

The reservoir is divided into many small tanks, cells or blocks to take into account reservoir heterogeneity. Computations using material

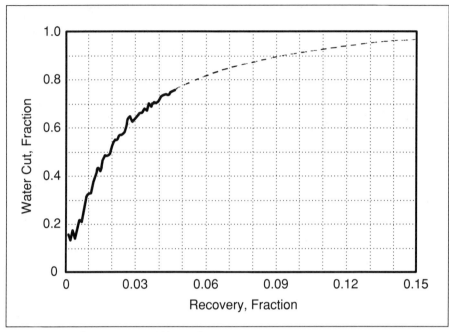

Figure 8-13 Recovery vs. Water Cut

balance and fluid flow equations are carried out for oil, gas and water phases for each cell at discrete time steps, starting with the initial time.

In general, reservoir simulation process can be divided into three main phases:[1]

1. Input data gathering
2. History matching
3. Performance prediction

Input Data

Input data for a black oil simulation generally consists of:

- General data for the entire reservoir – dimensions, grid definition, number of layers, original reservoir pressure, initial water-oil and gas-oil contacts. These data are obtained from base maps, log and core analyses and well pressure tests

- Rock and fluid data – Relative permeabilities, capillary pressures, rock compressibilities, and PVT data which are obtained from laboratory tests or correlations
- Grid data – geological data including elevations, gross and net thickness, permeabilities, porosities and initial fluid saturations. These data are obtained from well log and core analyses, and well pressure and well productivity tests
- Production/injection and well data – oil, water and gas production or injection history and future production and injection schedule for each well, well location, productivity index, skin factor and perforation intervals for each well

Gathering the needed data which can be very time consuming and expensive requires integrated team efforts involving geoscientists and engineers. Ascertaining the reliability of the available data and information is vital for the successful reservoir modeling.

History Matching

History matching of past production and pressure performance consists of adjusting the reservoir parameters of a model until the simulated performance matches the observed or historical behavior. This is a necessary step before the prediction phase because the accuracy of a prediction can be no better than the accuracy of the history match. However, it must be recognized that history matches are not unique. The reservoir parameters which may be adjusted must be identified and the degree of adjustment determined. Some reservoir data are known with a greater degree of accuracy than others. It is usually assumed, for example, that the fluid properties are valid, provided careful laboratory measurements were made. On the other hand, reservoir formation properties, i.e., porosity, permeability and capillary pressure, etc., are known only at the locations where the wells have penetrated the formation, and even these may be subject to error. In the inter well regions, the formation properties must be interpolated from geological and petrophysical correlations. Thus, if the well values are not precise, then the results of the simulation may also be inaccurate.

A step-wise history matching procedure is given below:

- Initialization verifying input of initial data
- Pressure matching by specifying production/injection for the wells and adjusting parameters affecting original hydrocarbon in place

- Saturation matching by adjusting relative permeability curves, vertical permeabilities, water-oil and gas-oil contacts, etc
- Well pressure matching by modifying productivity indices

Performance Prediction

Predicting future performance of a reservoir under existing operating conditions and/or some alternative development plan such as infill drilling, waterflood after primary, etc. is the final phase of a reservoir simulation study. The main objective is to determine the optimum operating condition in order to maximize economic recovery of hydrocarbon from the reservoir.

It should be noted that simulation studies can be useful even before production starts. This can be the case in new reservoirs. In this situation, reservoir simulation can be used to perform sensitivity studies to identify reservoir parameters which most influence hydrocarbon recovery. This can help in planning development strategies and identifying additional data which should be gathered.

The most comprehensive waterflood performance prediction tool is a reservoir simulator. Today, "black-oil model" simulators in main frame and better yet in PC are used extensively for waterflood performance prediction. These models permit inclusion of detailed reservoir description and laboratory-measured rock and fluid properties for more accurate predictions. One of the advantages is the ease of studying the effects of alternate operating strategies. Pattern type and size, infill drilling, effect of irregular patterns, injection rate scheduling, lifting capacities, zonal completions, etc. can be varied simply, and the effects observed. Many years of project life can be repeated under different operating strategies in a few seconds of high speed computer time.

Chapter 5 presents simulator analyses of various factors influencing 5-spot waterflood recovery efficiencies. Chapter 7 presents waterflood development strategy for a primary depleted reservoir using simulator.

A 40-acre, 5-spot pattern performance was simulated for partial primary depletion followed by waterflood. The rock and fluid properties were comparable to the example waterflood problem given in Craig's monograph.[9] The simulated results for homogeneous and layered cases with $(k_z = 0.1 \ k_x)$ and without vertical permeability are compared in Figure 8-14 with the classical pattern waterflood solutions. Even though the ultimate recoveries calculated by the various methods compare very favorably, the simulated performance shows considerably less total

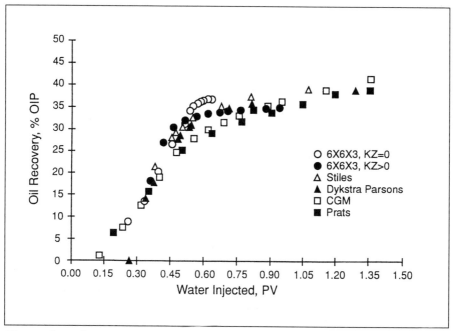

Figure 8-14 Comparison of Simulation and Classical Pattern Waterflood Solutions

water injection than the classical methods. Less water injection translates into shorter project life and favorable economics.

References

1. Thakur, C. G. "Waterflood Surveillance Techniques – Reservoir Management Approach," *JPT* (October 1991): 1181-1188.
2. Buckley, S. E. and Leverett, M. C. "Mechanisms of Fluid Displacement in Sands," *Trans. AIME* (1942) 146, 107-116.
3. Croes, G. A. and Schwarz, N. "Dimensionally Scaled Experiments and the Theories on the Water-Drive Process," *Trans. AIME* (1955) 204, 35-42.
4. Guthrie, R. K. and Greenberger, M. H. "The Use Of Multiple-Correlation Analyses for Interpreting Petroleum Engineering Data," Drill. and Prod. Prac., API (1955) 130-137.
5. Arps, J. J., et al. "A Statistical Study of Recovery Efficiency," API Bulletin D14 (1967).
6. Schauer, P. E. "Application of Empirical Data in Forecasting Waterflood Behavior," Paper 934-G presented at SPE 32nd Annual Fall Meeting, Dallas, TX, Oct. 6-9, 1957.

7. Guerrero, E. T. and Earlougher, R. C. "Analysis and Comparison of 5 Methods Used to Predict Waterflooding Reserves and Performance," Drill. and Prod. Prac., API (1961) 78-95.
8. Bush, J. L. and Halander, D. P. "Empirical Prediction for Recovery Rate in Waterflooding Depleted Sands," *JPT* (Sept., 1968) 933-943.
9. Craig, F. F., Jr. "The Reservoir Engineering Aspects of Waterflooding," SPE Monograph 3, Richardson, TX (1971).
10. Dykstra, H. and Parsons, R. L. "The Prediction of Oil Recovery by Waterflooding," Secondary Recovery of Oil in the United States, 2nd ed., API (1950) 160-174.
11. Stiles, W. E. "Use of Permeability Distribution in Waterflood Calculations," *Trans. AIME* (1949) 186, 9-13.
12. Prats, M., et al. "Prediction of Injection Rate and Production History for Multifluid Five-Spot Floods," *Trans. AIME* (1959) 216, 98-105.
13. Welge, H. J. "A Simplified Method for Computing Oil Recovery by Gas or Water Drive," *Trans. AIME* (1942) 146, 107-116.
14. Craig , F. F. , Jr. , Geffen, T. M. and Morse, R. A. "Oil Recovery Performance of Pattern Gas or Water Injection Operations From Model Tests," Trans., I (1955) 204, 7-15.
15. Higgins, R. V. and Leighton, A. J. "A Computer Method to Calculate Two-Phase Flow in Any Irregularly Bounded Porous Medium," I (June, 1962) 679-683.
16. Arps, J. J. "Estimation of Decline Curves," *Trans. AIME* (1945).
17. Arps, J. J. "Estimation of Primary Oil Reserves," *Trans. AIME* (1956) 207, 182-191.
18. Ershaghi, I. and Omoregie, O. "A Method for Extrapolation of Cut Vs. Recovery Curves," *JPT* (Feb. 1978) 203-204.

9

Waterflood Surveillance Techniques

This chapter describes waterflood surveillance techniques in the light of the reservoir management approach.[1,2] This approach considers a system consisting of reservoir characterization, fluids and their behavior in the reservoir, creation and operation of wells, and surface processing of the fluids. These are interrelated parts of a unified system. The function of reservoir management in waterflood surveillance is to provide facts, information, and knowledge necessary to control operations and to obtain the maximum possible economic recovery from a reservoir.

Guidelines for waterflood asset management should include information on (1) reservoir characterization, (2) estimation of pay areas containing recoverable oil, (3) analysis of pattern performance, (4) data gathering, (5) well testing and reservoir pressure monitoring, and (6) well information database.

Today, sufficient performance history is available that surveillance techniques can be documented in detail. This appendix highlights waterflooding in light of practical reservoir management practices. Case studies that illustrate the best surveillance practices are referenced.

Key Factors In Waterflooding Surveillance

Key monitoring points in the traditional waterflood cycle are described in Figure 9-l. In the past, attention was focused mainly on reservoir performance.[1] However, with the application of the reservoir management approach, it has become industry practice to include wells, facilities, water system, and operating conditions in surveillance programs.

It is important to consider the following items in the design and implementation of a comprehensive waterflood surveillance program (see Table 9-1).[2]

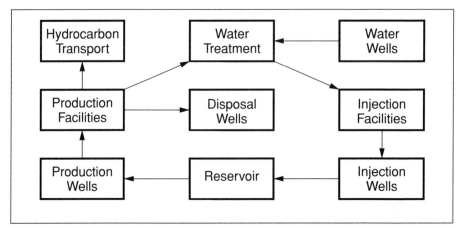

Figure 9-1 Waterflood Cycle (Copyright 1974, SPE, from JPT, December 1988[1])

Table 9-1 Waterflood Surveillance (Copyright 1991, SPE, from JPT, October 1991[2])

Reservoir	Wells	Facilities	Water System
Pressure	Perforations	Production/	Water quality
Rates	Production/injection	injection	Presence of corrosive
Volumes	logging	Monitoring	dissolved gases,
Cuts	Injected water in	equipment	minerals, bacterial
Fluid	target zone		growth, dissolved
samples	Tracer		solids, and suspended
Hall plots	Tagging fill		solids
Fluid drift	Cement integrity		Ion analysis
Pattern	Downhole		pH
balancing	equipment		Corrosivity
Pattern	Wellbore fractures		Oil content
realignment	Formation damage		Iron sulfide
	Perforation plugging		On-site or laboratory
			analysis
	Pumped-off		Data gathering on
	condition		source
			and injection wells and
			injection system

- Accurate and detailed reservoir description
- Reservoir performance and ways to estimate sweep efficiency and oil recovery at various stages of depletion
- Injection/production wells and their rates, pressures, and fluid profiles

- Water quality and treating
- Maintenance and performance of facilities
- Monthly comparison of actual and theoretical performance to monitor waterflood behavior and effectiveness
- Reservoir-management information system and performance control (accurate per-well performance data)
- Diagnosis of existing/potential problems and their solutions
- Economic surveillance

Reservoir Characterization and Performance Monitoring

Physical characteristics of the reservoir. Reservoir characteristics must be defined: permeability, porosity, thickness, areal and vertical variations, areal and vertical distributions of oil saturation, gas/oil and oil/water contacts, anisotropy (oriented fracture system or directional permeability), *in-situ* stress, reservoir continuity, vertical flow conductivity, and portion of pay containing the bulk of recoverable oil. To manage a waterflood accurately, detailed knowledge of the reservoir architecture also is necessary. Figures 3-14 and 9-2 show some examples of geological characterization, involving changing geological concepts and zonation.[3,4]

Primary performance. Wells indicating relatively high cumulative production may indicate high permeability and porosity, higher pay-zone thickness, or another pay zone. On the other hand, wells indicating relatively low cumulative production may indicate poor mechanical condition, wellbore skin damage, or isolated pay intervals.

Production curves. Percent of oil cut in the produced stream (log scale) vs. cumulative recovery during secondary performance may result in an estimate of future recovery or may indicate improvement in the waterflood performance as a result of more uniform injection profile. Figure 8-1 illustrates the performance curve of a typical successful waterflood, and Figure 9-3 illustrates various examples of waterfloods.[2]

Gas-oil-ratio. Decreasing gas-oil-ratios indicate that fluid fill-up is being achieved. Increasing gas-oil-ratios indicate that voidage is not replaced by injection.

Flood front map. This pictorial display shows the location of various flood fronts. The maps, often called "bubble maps," allow visual dif-

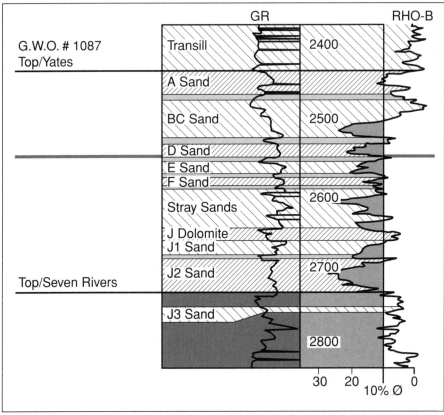

Figure 9-2 Type Log for North Ward Estes Field (Copyright 1990, SPE, from paper 20748[4])

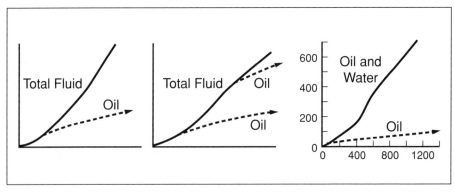

Figure 9-3 Cumulative Injection vs. Cumulative Total Fluid and Cumulative Oil (Copyright 1991, SPE, from JPT, October 1991[2])

ferentiation between areas of the reservoirs that have and have not been swept by injected water.[5] Before fill-up, Equations 9-l and 9-2 can be applied to estimate the outer radius of the banked oil and the water-bank radius.

$$r_{ob} = \left[\frac{5.615\ I_{cw} E}{\pi \phi h S_g} \right]^{1/2}$$
(9-1)

where:

r_{ob} = outer radius of the banked oil, ft
I_{cw} = cumulative water injected, bbl
S_g = gas saturation at start of injection, fraction
E = layer injection efficiency (fraction of water volume that enters the layer where effective waterflood is taking place)
h = thickness, ft

$$r_{wb} = r_{ob} \left[\frac{S_g}{\bar{S}_{wbt} - S_{iw}} \right]^{1/2}$$
(9-2)

where:

r_{wb} = water bank radius, ft
$\bar{S}_{wbt}$ = average water saturation behind front, fraction
S_{iw} = connate water saturation, fraction

If zones are correlative from well to well and if limited vertical communications exist, then the bubble map can be drawn for each zone. The bubble map can be used to identify areas that are not flooded and areas with infill drilling opportunities.

X-plot. Because extrapolation of past performance on the graph of water cut vs. cumulative oil is often complicated, a method was devised to plot recovery factor vs. X that yielded a straight line.[6,7] X was defined as:

$$X = [ln\ (1/f_w - 1)] - 1/f_w$$
(9-3)

where: f_w = fractional water cut.

This method is more general than the conventional plot of water cut vs. cumulative oil. Both of these methods are more applicable when the water cut exceeds 0.75.

Hall plot[8]. This technique, used to analyze injection-well data, is based on a plot of cumulative pressure vs. cumulative injection. It can provide a wealth of information regarding the characteristics of an injection well, as shown in Figure 9-4.

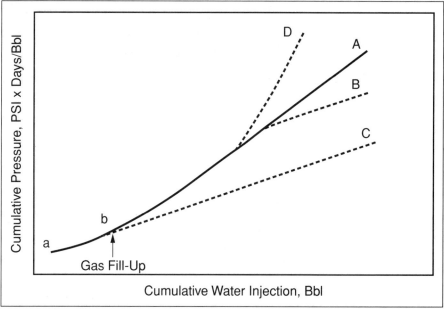

Figure 9-4 Typical Hall Plot for Various Conditions (Copyright 1991, SPE, from JPT, October 1991[2])

Early in the life of an injection well, the water-zone radius will increase with time, causing the slope to concave upward, as shown by Segment ab in Figure 9-4. After fillup, Line bA indicates stable or normal injection. An increasing slope that is concave upward generally indicates a positive skin or poor water quality (Line D). Similar slopes may occur if a well treatment is designed to improve effective volumetric sweep. In this case, however, the slope will first increase and then stay constant. Line B indicates a decreasing slope, which indicates negative skin or injection above parting pressure. The injection under the latter condition can be verified by running step-rate tests. A very low slope value, as shown by Line bC, is an indication of possible channeling or out-of-zone injection.

Controlled waterflood. Maximum profit and recovery would be realized if all wells reached the flood-out point simultaneously. This means producing the largest oil volumes from the wells draining the largest pore volumes. This scenario will result in minimum life with minimum operating expense while realizing maximum oil recovery. Note that if there is a large variation in pore volumes, this task is difficult because each well is allocated a production/injection rate on the basis of pore volume fractions.

Pattern balancing. Minimizing oil migration across pattern boundaries improves the capture of the mobilized oil and reduces the volume of recycled water. Pattern balancing generally increases sweep efficiency. In addition, realignment of flood patterns in conjunction with pattern balancing provides more opportunities to increase oil recovery. Simple reservoir modeling work can be helpful. For example, the modeling work can identify an unswept area and what kind of improvement in sweep can be obtained by changing producer/injector configuration.

Produced water analysis. Injected-water breakthrough can be detected by monitoring the chloride content of the produced water if there is a significant difference in the salinities.

Injection profile surveys. Periodic surveys of injection-well fluid entry profiles can detect formation plugging, injection out of the target zone, thief zones, and underinjected zones. Allocation of injection volumes with data obtained from the profile surveys allows tracking of waterflood histories of each zone.

Wells

Problem areas. Formation plugging, injection out of the target zone, and nonuniform injection profile caused by stratification are all problem areas. They cause major problems in waterflood operations and low vertical sweep efficiency. Thin, high permeability layers serve as highly conductive streaks for the injected water.

Well completion. Condition of the casing and/or cement bond plays an important role in waterflood surveillance. Because of poor cement, water flow can occur behind the casing. Also, openhole injectors and producers and fractured wells with large volume treatments are not generally desirable. The latter condition may sometimes have a significant negative effect on sweep efficiency. Note that these conditions do not preclude a successful waterflood, but they require more concentrated efforts in surveillance.

Injection well testing. These tests are conducted to optimize waterflood performance by maximizing pressure differential, minimizing skin damage, ensuring proper distribution of water, and monitoring the extent of fracturing.

Quality of producers. If producers are converted to injectors, care should be taken to avoid conversion of all poor producers because generally poor producers make poor injectors.

Converting producers. Producers are converted and high gas producers are shut in to accelerate fill-up time.

Backpressure. If the producing wells are not pumped off, a backpressure is applied to cause crossflow. As a result, the low-pressure zones may not produce.

Changing injection profiles. This can be done with selective injection equipment, selective perforating, low-pressure squeeze cementing, acidizing, and thief zone blockage through polymer treatments.

Regular cleanouts. Regular cleanouts of injectors are necessary, especially if they become plugged with time because of unfiltered water. A Hall-plot analysis may provide some guidance regarding the well cleanout necessities (Figure 9-4). Regular cleanouts not only increase injectivity, but they also improve the injection profile because the low permeability zones are the first ones to plug, which causes the profile to become less than desirable. Of course, the decision on the frequency of well cleanouts should be made based upon economics.

Completion and workover techniques. Selection of completion and workover fluids, perforating, and perforation cleaning should be carefully made. This may maximize the completion efficiency of the injection and production wells.

Flow regulation. Proper utilization of surface and downhole regulator and single/dual-string injector should be made.

Profile control. Polymer, cementing, chemical, and microbial methods may assist in controlling profiles for improving vertical sweep.

Facilities/Operations

The literature on waterflood surveillance is generally aimed at reservoir performance. Overall project success, however, is often critically affected by daily field operations. While reservoir engineers and geologists play a very important role in reservoir performance and waterflood optimization, facilities/operations staff are concerned with daily management of field operations, information collection, and diagnosis of potential or existing problems (mechanical, electrical, or chemical).

Surface equipment considerations should include a surface gathering and storage system, injection pumps, water distribution systems, metering, water treatment and filtering system, oil/water separation, corrosion and scale, plant and equipment sizing, and handling of separated waste products.

Water-Quality Maintenance

If water quality is not maintained, higher injection pressures are required to sustain the desired injection rates. Also, corrosion problems increase with time when lower-quality water is used. It is important to protect the injection system against corrosion to preserve its physical integrity and to prevent the generation of corrosion products.

Ideally, the water quality should be such that the reservoir does not plug and injectivity is not lost during the life of the flood. However, cost considerations often prohibit the use of such high-quality water. The expense of obtaining and preserving good-quality water must be balanced against the loss of income incurred as a result of decreased oil recovery and increased workover and remedial operations requirements.

Questions are often asked about the determination of acceptable water quality. Tighter formations require better quality water. Sometimes poor quality water can be injected above parting pressures, but injection through fractures could reduce sweep efficiency.

Although it is impossible to predict quantitatively, the minimum water quality required for injection water into a given formation, some attempts have been made and documented in the literature to define injection water-quality requirements from on-site testing.[9] Table 9-2 and Figure 9-5 describe other considerations regarding water systems.[2]

Table 9-2 Water System (Copyright 1991, SPE, from JPT, October 1991[2])

Water source (produced, source well, separated)
Water-quality requirements:
- source water—produced-water compatibility
- injection water–reservoir/rock interaction (clay swelling)
- dispersed oil
- corrosion
- scale
- bacteria (sulfate reducing, oxidize soluble iron in water, produce organic acids)
- marine organisms
- pH control
- corrosive dissolved gases (CO_2, H_2S, O_2)
- total dissolved and suspended solids (iron content, barium sulfate)
- corrosion inhibitors (not sufficiently soluble)
- scale inhibition
- closed vs. open injection facilities
- treatment program to ensure acceptable water for formation and to minimize corrosion

Other important considerations:
- oil/water separation
- filtration (gathering station, treatment plan, types of filters, wellhead filters and strainers)
- waste treatment
- water-supply wells (solids, corrosion products, bacteria)
- surface water (oxygen, bacteria, marine organisms, suspended inorganic solids)

It is interesting to note that incompatible barium and sulfate waters were injected into the Baylor County Waterflood Unit No. 1.[10] Produced and makeup waters were not mixed; instead, they were injected through two separate systems and into separate wells. No problems were encountered through mixing and precipitation in the reservoir, nor were any problems in the producing system experienced.

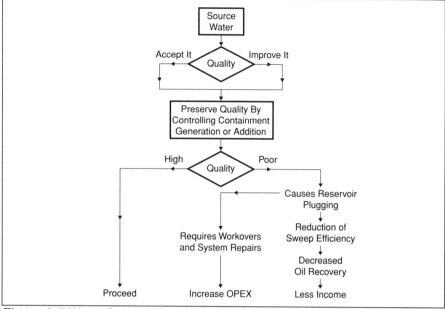

Figure 9-5 Water System (Copyright 1991, SPE, from JPT, October 1991[2]

Monitoring

Reservoir

- Pressure (portable test equipment, fluid-level testing; repeat formation, buildup/fall-off, and step-rate tests; fieldwide pressure surveys to determine pressure gradient for use on balancing injection/production rates)
- Rate (oil, water, gas, water-cut, GOR, well testing—production rates)
- Pattern balancing (voidage control, areal/vertical sweep efficiency using stream-tube models)
- Waterflood pattern realignment
- Observation/monitoring wells
- Reservoir sweep and bypassed oil
- Fracture communications
- Thief zones and channels
- PV injected
- Gravity underriding and fingering/coning

Wells

- Production/injection logging (openhole/cased hole, temperature/spinner/tracer)
- Injected water in target zone
- Hall plots (well plugging/stimulation)
- Tracer (single well/interwell)
- Tagging fill
- Cement integrity
- Downhole equipment
- Surface equipment
- Wellbore fractures
- Formation damage/perforation plugging
- Pumped-off condition
- Corrosion/scale-inhibition residuals

Facilities

- Production/injection
- Monitoring equipment and maintenance

Water System

- Presence of corrosive dissolved gases (CO_2, H_2S, O_2); minerals; bacterial growth; dissolved solids; suspended solids, concentration and composition; ion analysis; pH
- Corrosivity (corrosion coupons and corrosion rate monitoring), oil content (dispersed or emulsified oil in water), and iron sulfide
- On-site or laboratory analysis
- Data gathering at the water-source well, water-injection wells, and several points in the injection system

Case Histories

Means San Andres Unit

A comprehensive surveillance program used at the Means San Andres Unit was documented by Stiles.[11] A detailed surveillance program was developed and implemented in 1975. It included monitoring production (oil, gas, and water) and water injection, controlling injection pressures with step-rate tests, pattern balancing with computer balance program, running injection profiles to ensure optimum distribution, selecting specific production profiles, and choosing fluid levels to ensure pumpoff of producing wells.

The following were implemented during tertiary recovery (water-alternating-gas injection), but they also apply to waterflood surveillance.

- Areal flood balancing (optimizing the arrival of flood fronts at producers) performed by annual pressure-falloff tests in each injection and computer balancing programs
- Production/injection monitoring
- Data acquisition and monitoring
- Pattern performance monitoring to maximize oil recovery and flood efficiency by evaluating and optimizing the performance of each pattern
- Optimization (it must be dynamic and sensitive to changes in performance, technology, and economics)
- Vertical conformance monitoring to optimize vertical sweep efficiency while minimizing out-of-zone injection. Several cross-sections were constructed for each pattern to ensure completions in all the floodable pay. Annual profiles were run on all injection wells. For each profile, casing or packer leaks were identified, out-of-zone injection was identified, and zonal injection from profile was compared with porosity-feet profiles

The main objective of an injection survey is to provide a means of monitoring the injection water so that efforts can be made to ensure that injection rates conform with zonal porosity-thickness. These efforts have paid substantial dividends in increased vertical sweep and ultimate recovery.

South Hobbs Unit

Production at the South Hobbs Unit increased almost 100% within a year.[12] The reason for boosted performance was an aggressive program of well surveillance, general record keeping, and remedial action. Five operational efforts had positive effects on production.

- Lift capacity of a number of wells was increased so that a pumped-off condition could be maintained
- Operating pressures in the satellite battery separators were reduced, thereby reducing backpressure through the flow-lines back to the well
- Adverse effects of scale accumulation were decreased by remedial and preventive measures.
- Injection pressures just under the parting pressure were maintained
- Tracer surveys were run to ensure that fluid is entering into the proper zones in the right amounts

West Yellow Creek Field

This case study describes the importance of a thorough, well-organized reservoir surveillance for the West Yellow Creek field.[13] This effort involved many activities, including pressure-falloff tests, a computerized flood balancing program, and a produced-water sampling program.

Ventura Field

Schneider described the role of geological factors on the design and surveillance of waterfloods in the structurally complex reservoirs in the Ventura Field, California.[14] Geologic factors strongly influenced the profiles of injection wells and the responses of producing wells. The waterflood was monitored to determine the cause of injection anomalies and to predict their effect on waterflood response.

Jay/Little Escambia Creek Flood

The application of reservoir management techniques was key to the success of this waterflood.[15-17] Surveillance information and reservoir description data provided new insights into water movement and zonal depletion. Operating decisions based on these data proved to be highly profitable.

Surveillance was used for both the vertical and horizontal conformances. Cased-hole logging, pressure-buildup and production tests, and permeability data from core analysis were used for the vertical conformance surveillance; radioactive tracers, reservoir pressure data, and interference tests were used for the areal surveillance.

To achieve vertical conformance, injection wells were acid-frac-ture-treated in multiple stages to create connecting vertical fracture systems. Temperature surveys, noise logs, and flowmeters were used for the vertical conformance surveillance. The entire section in the producing wells was opened without acid fracturing to maintain the flexibility of future water production. Flowmeter/gradiomanometer surveys, pressure buildups along with core analysis data, noise logs, and gamma ray logs were used for monitoring. In addition, pulsed-neutron-capture logs tracked edgewater encroachment.

Radioactive tracer data provided a means of determining the source of water breakthrough, which was later confirmed by the interference test performed between the producer and the suspect injection well. On the basis of these results, injection rates were adjusted to minimize trapped oil behind the water fronts.

Wasson Denver Unit

Ghauri, et al. described several innovative techniques to increase this unit's production rates and reserves, including novel geological concepts (Figure 3-14), major modifications in flood design, infill drilling, and careful surveillance.[3,18,19] The waterflood surveillance incorporated such common techniques as computer-generated analyses of production/injection data, water-bank radii or bubble maps, pressure contour maps, artificial lift monitoring, and specific items like careful monitoring of the relationship between reservoir withdrawals and the water-injection rate. The latter was monitored on both a unit and individual battery basis.

Based upon an energy balance of injection and withdrawals,

$$B_w i_w = B_o q_o + B_g (R - R_s) q_o + B_w q_{w'} \qquad (9\text{-}4)$$

where

B_w	=	water FVF, RB/STB
i_w	=	water injection rate, STB/D
B_o	=	oil FVF, RB/STB
q_o	=	oil production rate, SIB/D
B_g	=	gas FVF, RB/scf
R	=	producing GOR, scf/STB
R_s	=	solution GOR, scf/STB
q_w	=	water production rate, STB/D

or,

$$q_o = \frac{B_w (i_w - q_w)}{B_o + B_g (R - R_s)} \qquad (9\text{-}5)$$

With 800-psi PVT data and injection and production rates of 416,000 and 70,000 STB/D water, the oil rate for producing GORs of 700 and 750 scf/STB are 148,000 and 138,000 STB/D, respectively (B_o = 1.213 RB/STB, B_g = 0.003125 RB/scf). The rate model was history matched with actual performance in individual battery areas and then used to investi-gate the effects of changes in operating policy.

A significant effort was also made to improve the vertical sweep efficiency in both existing and new water-injection wells. Cemented liners were installed in openhole producers that were converted to injection, and the zones to be flooded were selectively perforated. All new producers and injectors were cased through the zones of interest and selectively perforated rather than completed openhole, which had been practiced before. Treating pressures during acid stimulation jobs were kept under formation fracturing pressures to maintain zonal isolation, and injection rates below fracturing pressure were maintained.

Other items of interest in surveillance and monitoring are described in Appendix E. They include:

1.   determination of S_{or}/ROS (residual/remaining oil saturation)
2.   injection and production logging
3.   interwell tracers

References

1.   Talash, A. W. "An Overview of Waterflood Surveillance and Monitoring," *JPT* (December 1988): 1539-43.
2.   Thakur, G. C. "Waterflood Surveillance Techniques—A Reservoir Management Approach," *JPT* (October 1991): 1180-1188.
3.   Ghauri, W. K., A. F. Osborne, and W. L. Magnuson. "Changing Concepts in Carbonate Waterflooding—West Texas Denver Unit Project—An Illustrative Example," *JPT* (June 1974): 595-606.
4.   Thakur, G. C. "Implementation of a Reservoir Management Program." Paper SPE 20748 presented at the 1990 SPE Annual Technical Conference and Exhibition, New Orleans, September 23-26.
5.   Staggs, H. M. "An Objective Approach to Analyzing Waterflood Performance." Paper presented at the Southwestern Petroleum Short Course, Lubbock, Texas, 1980.
6.   Ershaghi, I. and O. Omoregie. "A Method for Extrapolation of Cut vs. Recovery Curves," *JPT* (February 1978): 203-204.

7. Ershaghi, I. and D. Abdassah. "A Prediction Technique for Immiscible Processes Using Field Performance Data," *JPT* (April 1984): 664-670.
8. Hall, H. N. "How to Analyze Waterflood Injection Well Performance," *World Oil* (October 1963): 128–130.
9. McCune, C. C. "On-Site Testing to Define Injection-Water Quality Requirements," *JPT* (January 1977): 17-24.
10. Roebuck, I.F., Jr. and L. L. Crain. "Water Flooding a High Water-Cut Strawn Sand Reservoir," *JPT* (August 1964): 845-50.
11. Stiles, L. H. "Reservoir Management in the Means San Andres Unit." Paper SPE 20751 presented at the SPE Annual Technical Conference and Exhibition, New Orleans, Sept. 23-26, 1990.
12. Sloat, B.F. "Measuring Engineering Oil Recovery," *JPT* (January 1991): 8-13.
13. Gordon, S. P. and O. K. Owen. "Surveillance and Performance of an Existing Polymer Flood: A Case History of West Yellow Creek." Paper SPE 8202 presented at the SPE Annual Technical Conference and Exhibition, Las Vegas, Sept. 23-26, 1979.
14. Schneider, J. J. "Geologic Factors in the Design and Surveillance of Water-floods in the Thick Structurally Complex Reservoirs in the Ventura Field, California." Paper SPE 4049 presented at the SPE Annual Meeting, San Antonio, Oct. 8-11, 1972.
15. Langston, E. P. and J. A. Shirer. "Performance of Jay/LEC Fields Unit Under Mature Waterflood and Early Tertiary Operations," *JPT* (February 1985): 261-68.
16. Langston, E. P., J. A. Shirer, and D. E. Nelson. "Innovative Reservoir Management—Key to Highly Successful Jay/LEC Waterflood," *JPT* (May 1981): 783-91.
17. Shirer, J. A. "Jay/LEC Waterflood Pattern Performs Successfully." Paper SPE 5534 presented at the SPE Annual Technical Conference and Exhibition, Dallas, Sept. 28-Oct. 1, 1975.
18. Ghauri, W. K "Innovative Engineering Boosts Wasson Denver Unit Reserves," *Pet. Eng.* (December 1974): 26-34.
19. Ghauri, W. K "Production Technology Experience in a Large Carbonate Waterflood, Denver Unit, Wasson San Andres Field," *JPT* (September 1980): 1493-1502.

10

Field Operations

This chapter discusses various aspects of field operations related to waterflood asset management, including water system, compatibility and treatment; water injection and production control; conversion versus newly drilled wells, and methods of increasing injectivity. It should be emphasized that the success of a waterflood depends upon how well field operations are managed. The operations must be closely integrated with geological and engineering activities.

Water System, Compatibility, and Treatment

Water Quality/Handling

Maintenance of water system, compatibility between various fluids, and treatment to achieve desired fluid characteristics are discussed in this section.

If water quality is not maintained, higher injection pressures are required to sustain desired injection rates. Also, corrosion problems increase with time when lower-quality water is used. It is important to protect the injection system against corrosion to preserve its physical and chemical integrity and to prevent the generation of corrosion products.

This section discusses water treating facilities and surface and downhole equipment for injection or disposal.[1,2,3] The field operations generally are concerned with:

Reducing operating costs. Properly designed, constructed, operated, and monitored facilities for a waterflood reduce the costs of treating source water for injection and produced water for disposal. Properly treated injection water reduces the costs associated with workovers and stimulation of injection wells. Adequate corrosion and scale control measures extend the operating life of equipment, decrease maintenance and operating costs, and reduce lost production. Control of bacterial activity can maintain sweet production, minimize the production of toxic and corrosive gases, maintain product quality and value, and minimize the costs associated with more expensive, sour service operations. Water quality of any degree can be achieved; the goal in cost-effective treating is to balance the cost of treating with the remedial costs of not treating.

Increasing oil production. Proper water treating, surface, and downhole equipment assists in maintaining design water injection rates and profiles to maximize oil production rate and ultimate recovery.

Meet environmental regulations. The goal here is to go beyond just meeting environmental regulations and to exceed set regulations. To participate in and to be a leader is setting requirements to adequately protect the environment while maintaining appropriate economic development for the employees, for the company, and for the country. Properly designed and operated separation and treating facilities reduce the amount of oil discharged with the produced water and the impact of that oil upon the environment. Another example—adequate corrosion control procedures reduce the occurrence of leaks that can cause oil and produced water spills that harm people and the environment.

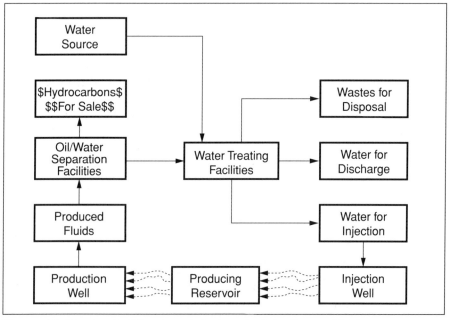

Figure 10-1 Integrated Treatment System for Injection or Discharge

Treatment system for injection/discharge.

The Integrated Treatment System (Figures 10-1 and 10-2) approach is considered with water treatment as an integral part of the production-separation-discharge-injection system. The separation-treating facilities are designed, constructed and operated with consideration given to the characteristics of the produced fluids, the source water, and the reservoir properties that will set the water quality requirements for injection and the environmental regulations and constraints applied to discharges. Comparison of the quality requirements for injection and/or discharges with the properties of the produced water and/or the source water will indicate the types and degrees of treatment for each water. In turn, chemical additions to the produced fluids and operation of the separation facilities will affect the quality of the produced water from these facilities and any additional treatment necessary to meet injection or discharge requirements. Finally, the water that is injected for additional oil recovery will eventually pass through the reservoir and reappear with the produced fluids. Or, the water that is discharged may reappear in the source water. Thus, every part of the producing-separation-treating-injection-discharge-reservoir system is linked.

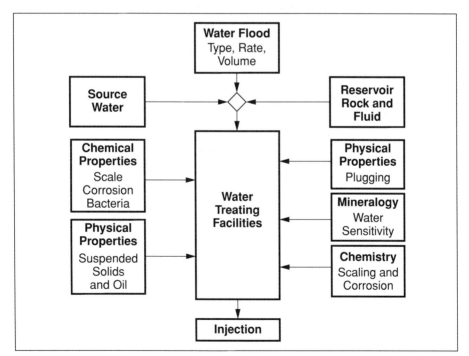

Figure 10-2 Strategy for Selection of Water Source and Treatment

Process steps for selection of water treatment equipment (see Table 10-1). Three primary variables play a role in designing and operating water treating equipment: (1) the type of waterflood and the properties of the injection formation; (2) the nature of the produced water and possible methods for its disposal; and (3) identification and characterization of potential water sources for the flood. The type of waterflood in terms of volume, rate and timing will have a marked influence on the selection of the water source. Reservoir characteristics and environmental constraints will set water quality requirements for injection and discharge, respectively. The properties of the produced water and any additional source water needed and the water quality requirements will dictate what treatment processes or options are suitable to convert or treat the produced water and the source water to the required quality. Usually several types of equipment can be used to treat a particular component or to achieve the required quality. Selection of the most cost effective process within the constraints of cost, safety and environment is essential. Once the processes and equipment are selected, designed and constructed, they must be operated efficiently and the performance monitored to assure that they are achieving the required water quality.

Table 10-1 Process Steps for Water Treating Equipment

- Characterization of Waterflood and Reservoir
- Identification of Produced Water and Disposal Methods
- Identification and Characterization of Alternate Water Sources
- Specification of Water Quality Requirements for Injection and Discharge
- Selection of Treating Operations to Achieve the Required Water Quality
- Design, Construct and Operate the Process Equipment
- Monitor Equipment Performance

Water quality requirements (see Table 10-2).

Water quality can be defined as the sum total of all the chemical, physical and microbiological properties required for a water to be suitable for a specified application. Water quality requirements are application- and site-specific. They may be set by the application or they may be determined by some regulatory agency. Water quality specifications may cover a wide variety of properties such as temperature, color or turbidity; dissolved components including inorganic and organic solids, liquids and gases; suspended material including solids from corrosion or scale forming reactions or dispersed oil from separation facilities; or even microscopic biological components that can produce fouling, plugging, corrosive, and toxic materials. Each application will have its own set of water quality requirements. The challenge is to determine the quality requirements for a particular application, to devise a treatment process to achieve the required quality, to operate the equipment is a cost effective and environmentally acceptable manner, and to monitor over all plant operation to assure the required water quality is being achieved.

Table 10-2 Water Quality Requirements for Injection

- Adequate Volume and Rate
- Low Plugging and Fouling Characteristics
- Noncorrosive
- Compatible with Formation Rock
- Compatible with Formation Fluids
- Nonscaling
- Sterile
- Not Harmful to Personnel or the Environment

The ideal injection water would be of such a quality as to have no detrimental interactions with or effects upon the treating equipment, the fluids and rocks of the producing reservoir, or the rate, amount and properties of the recovered oil. In general, the ideal injection water would have the following characteristics: (1) no dissolved solids or gases that could cause or promote corrosion or scale formation; (2) no suspended solids or dispersed oil that could foul the surfaces of treating equipment and connecting piping or plug injection wells and the formation; (3) no adverse reactions with the reservoir rock or reservoir fluids; (4) sterile and incapable of supporting bacterial and microbiological growth; (5) no materials that would affect oil properties to the extent that they would degrade the quality or amount of oil produced or increase problems with oil treating and oil-water separations; and (6) no substances that would be harmful to personnel or to the environment that might come in contact with the water. Additionally the injection water must be available in volumes and rates as specified by the asset team.

The ideal injection water does not exist. In all cases, water from some source must be treated to approach the desired requirements. The challenges in design and operation of water treating equipment for injection are: (1) to determine the minimum acceptable quality requirements for a specific waterflood; (2) to design and construct minimal, low cost, safe and effective treating equipment and processes to achieve the required quality; and (3) to operate the facilities in a safe and cost-effective manner while still producing injection water of the required quality.

Specification of requirements for the discharge of produced water and other solid and liquid wastes is more involved than those for injection. The specifications are usually set by a regulatory agency. It, in addition to setting discharge standards, will set such requirements as sampling schedules and locations, required analysis procedures, and reporting requirements. Additionally, an operator may set discharge standards over and above those of the regulatory agency to insure that operations are in harmony with and not detrimental to the local population, economy, or environment. Consequently, compliance requirements for discharges will be very site specific and will vary with the type and the amount of discharge as well as the nature and the species with the receiving environment. Most discharge permits will include some restrictions on the total amount and the concentration of dispersed and sometimes soluble oil that can be discharged. Other items that might be included for specific sites and discharges are: temperature, suspended solids, turbidity, biological or chemical oxygen demand, dissolved oxygen, pH, salinity (total dissolved solids), specific chemical compounds, and toxicity.

Compatibility. Two types of water compatibility exist. The first type concerns compatibility between the water that is injected into the reservoir and the water that is already present within the pores of the formation rock. The second type of compatibility is concerned with possible detrimental reactions between the injection water and the reservoir rock itself.

Suspended solids and dispersed oil. Suspended solids and dispersed oil are the principal quality parameters for injection water. They are responsible for most of the plugging and injection losses that occur. Most of the capital and operating costs in injection water treatment operations are associated with the removal and control of these two components.

Corrosion control. Corrosion control will consider reactions between the waters and the metal structures involved in the production/treating/injection system. Primary concern will be on how corrosion reactions affect water quality and injection rates. The corrosion reactions and their effect upon the structural materials of the system will receive less emphasis.

Scale inhibition. Scale inhibition is concerned with detecting and controlling chemical precipitation and deposition reactions that might occur from chemical and physical changes happening in the water or from mixing of two incompatible waters.

Microbiological effects. Bacterial and microbial effects will consider such items as plugging of wells and fouling of surfaces by bacteria and biomass, microbial production of toxic and corrosive gases such as hydrogen sulfide and carbon dioxide, and microbial reservoir souring.

Regulated soluble components. The dissolved materials are primarily a concern in water that is discharged rather than injected. These materials become important when the type and amounts that can be discharged are set by environmental regulations. Materials that might be included are soluble organic materials that register as "oil," salinity, specific compounds such as phenol, ammonia, and toxic materials.

Identification of Potential Water Sources

The volume, rate, and timing of the waterflood identified previously will assist in selecting potential water sources. The source must be able to supply the required volumes and rates during the course of the waterflood. Identification of potential water supplies may also be influenced by positive and negative environmental constraints. For example, there may be influence to reduce environmental impacts by reinjecting the produced water rather than discharging it.

There are a number of potential water sources. Listing them in decreasing order of choice, they are (1) produced water from the reservoir that is to be flooded; (2) produced water from another reservoir (with similar properties to the one being flooded); (3) aquifer or source wells in nonproducing zones; (4) surface waters such as rivers, lakes or seawater; and (5) waste waters. Produced water from the reservoir that is being flooded is the preferred source because it usually requires the least treatment for reinjection and there is less possibility of detrimental reactions with reservoir rocks and fluids. Waste water is the least acceptable because treating costs to achieve the necessary injection water quality will usually be very high.

In some cases, a single source may not have adequate volume. Then, consideration of multiple sources is necessary. Treating each source separately and then using separate injection systems for each source invariably is the simpler solution and will present the fewest problems. Mixing source waters before treatment and injection usually leads to significant handling problems.

Figure 10-2 summarizes the strategy involved in matching the quality requirements of water for injection or discharge with the properties of the source water through appropriate processes in the water treating facilities. Thus, the type of waterflood in terms of injection rates and volumes is needed to screen the potential water sources. The properties of the reservoir rocks and fluids will dictate the required water quality for injection. Environmental regulations will set the requirements for discharge. Matching the required water qualities for injection or discharge with the properties of the source water and the produced water will suggest necessary treating operations to transform the waters. As just one example, the porosity and permeability of the reservoir will influence the amount and size of suspended solids that can be tolerated in injection water to maintain a suitable injection rate. Comparing these requirements with the suspended solids content of the source water will indicate if solids removal might be required and, if so, the type of solids removal process that might be suitable. If filtration is used for solids removal, the type of filter used will set the maximum permissible oil content in the feed to the filters.

Compatibility – Water/Rock and Water/Water Interactions and Formation Damage

Whenever there is a significant difference in either the composition or the concentration of the injected water and the formation water that has been in contact with reservoir rocks for long times, a potential for detrimental interactions exits. This is frequently the case when a source water, used for injection, is different from the produced water. We will consider two types of compatibility or reactions as they are related to water quality and water injection. The first is concerned with chemical and physical reactions and interactions between the injection water and the formation rock in the injection zones. These reactions occur because of differences in chemical composition and concentration between the injection water and the water with which the rock has been in equilibrium. The second type of compatibility is concerned with precipitation reactions that might occur between the water that is being injected into the formation and the water that is already in the formation—this may be the connate water or some other previously injected water. This type of interaction will be discussed later in the section on scale control.

There are two general mechanisms for formation damage or permeability impairment from injected water/formation rock interactions. The first mechanism is concerned with the formation and liberation of fines. These fines then move with the water flow until they become lodged in a pore throat. Processes involved in this mechanism include: (1) clay swelling, (2) clay deflocculation, (3) mica alteration, and (4) critical flow velocity. The second mechanism involves solution of formation minerals. In some cases, damage occurs when the dissolved minerals reprecipitate. In the other case, the solution of some minerals frees less soluble minerals that can move with the water flow and plug pore throats.

The key to maximizing injection efficiency (rates and profiles) is to identify the potential problem before hand and then to treat to prevent or control the damage. The less efficient procedure is to try to identify the damage mechanism after it has occurred and then to try remedial treatments. *Generally the best option is to be proactive rather than reactive.*

There are a number of test procedures available to detect and identify possible damage upon water injection. They include: (1) core flow tests (not to be confused with core flow tests to determine required water quality for injection); (2) mineral analysis of the formation rock; and (3) chemical analysis of the waters involved in the flood. The tests are usually run in the order listed. The rationale for this order is first to run the core flow tests to see if there is a problem, and then if there is, the mineralogical and chemical analysis can be used to determine the cause of the problem in the rock and water phases, respectively.

If several water sources are available, choosing the least damaging is the most economical. This is particularly true if produced water is available in sufficient volumes. Since the produced water came from the formation, it should be less damaging when reinjected into the same formation, particularly if care has been exercised to minimize chemical and physical changes in the water during production, separation, treatment, and injection.

Modifying the completion techniques and the injection procedures can be especially helpful when the damage mechanism is fines migration from high flow rates. Lower injection rates and completion techniques (higher perforation density, open hole completions, etc.) that increase the cross-sectional area for flow may reduce the velocity below the critical value, particularly around the injection wells where velocities are the highest.

Changing the water chemistry by the addition of divalent cations or potassium salts is usually not economical for the volumes involved in an injection project. Such treatments may be justified for low volume conditions such as drilling or completion fluids.

There are two variations available for protecting the formation rock around a wellbore prior to injection of a damaging water. First, inject a calcium chloride, potassium chloride, or ammonium chloride brine. Or second, inject a commercially-available clay stabilizer. The first treatment is less expensive but not as long lasting as the second.

Water Treatment for Injection and Discharge

There are two general types of water to be treated: produced water for injection or discharge and sources water for injection. We are going to break the discussion on water treatment into three parts: (1) Preliminary Treatment of Source Water for Injection, (2) Preliminary Treatment of Produced Water for Injection or Discharge, and (3) Common Water Treatment Processes for Injection or Discharge. In the first two sections we will discuss required operations that are specific to a particular water type. In the third section, we will discuss items common to both types of water.

Preliminary treatment of source water for injection. Here we will be concerned with those operations that are unique to treating source waters from surfaces or from shallow aquifers and we will be concerned only with treating for injection. Preliminary processes in this category include control of biological growth in the inlet system and removal of dissolved oxygen (deaeration or deoxygenation).

Preliminary treatment of produced water for injection or discharge. This section will be concerned with preliminary operations for produced water to prepare it for further treating processes for either injection or discharge. The discussion will cover basic oil-water separation process, also called oil dehydration and removal of oil from the separated produced water, also called oily water clean up, prior to other processes to further clean the produced water.

Common water treating processes for injection and discharge. We return to processes that would be needed for either produced water or source water for injection and produced water for discharge. Operations that will be considered include removal of suspended solids and soluble components (removal of solubles is usually only necessary for produced water discharges), scale inhibition, and control of corrosion and microbiological effects.

Next to produced water, surface water and seawater in particular is the second most popular water source for injection. In terms of volume, more seawater is used for waterfloods than any other water source. Seawater and all surface waters contain certain undesirable components that must be removed before the water is suitable for injection. These are: (1) bacteria and other biologically active substances, (2) dissolved oxygen, and (3) suspended solids. Biologically active material can grow and multiply to produce slimes and biofilms that foul surfaces of piping and vessels and that plug porous material such as media filters and the formation into which the water is injected. Dissolved oxygen, particularly in higher salinity waters such as seawater, is extremely aggressive and corrosive towards carbon steel. Carbon steel, because of its low cost and ease of working, is the usual material of choice for most oilfield systems. Finally, suspended solids can foul surfaces and plug porous materials. Suspended solids damage injection wells resulting in decreased injectivity and altered injection profiles. In this section, we will be concerned with removal of dissolved oxygen and control of biological activity. Suspended solids removal is common to source water and produced water treatment and will be covered in a later section on filtration.

In a typical seawater injection system, seawater is picked up at some distance below the surface and above the bottom to minimize the amount of suspended solids present. Strainers are used on the suction side of the seawater lift pump to keep out the larger size material. Chlorination is almost universally used to minimize fouling by biological growth and bacteria-generated biofilms. Liquid chlorine is extremely toxic and difficult to handle. For safety reasons, many offshore treating facilities generate chlorine on site. Chlorine can be formed by passing an electric current through seawater in an electrochlorinator. Chlorine dosages are easily adjusted by varying the strength of the electric current. When chlorine is added to water, it forms hydrochloric and hypochlorous acids according to the following equation:

$$Cl_2 + H_2O \longrightarrow HCl + HOCl$$

The hypochlorous acid in turn ionizes to form the hypochlorite ion:

$$HOCl \longleftrightarrow H^+ + OCl$$

The hypochlorous acid, HOCl, is the active material to control biological and bacterial growth. Because hypochlorous acid is a weak acid, its ionization is dependent upon the solution pH. At high pH typical of seawater, most of the chlorine appears as the relatively inactive hypochlorite ion. In practice, the low effectiveness of the hypochlorite ion is counteracted by using higher chlorine dosages. Chlorine and hypochlorous acid, its form when dissolved in water, is a very effective biocide. Total chlorine residuals < 1 ppm are usually sufficient for biological control. Chlorine is also a strong oxidizing agent. Oxidizing agent are corrosive to carbon steel. Care must be taken to keep the chlorine residual or "free" chlorine at a level sufficient for biological effectiveness but not so high that corrosion rates become excessive.

Dissolved oxygen, particularly in high salinity waters such as seawater, is very aggressive towards carbon steel. Carbon steel, because of its lower cost and ease of working, is the material of choice for most oilfield equipment. Most corrosion inhibitors have limited effectiveness in the presence of oxygen. The preferred practice when dealing with surface waters is to reduce the dissolved oxygen levels to below 10-20 parts per billion (ppb). Seawater can contain as much as 6-10 parts per million (ppm) of dissolved oxygen.

Particulates – dispersed oil and suspended solids. Particulates are defined as disperse (free) oil or suspended solids. They are present in separated produced waters. Dispersed oil is usually not found in source waters if the pick up point for the source water is remote from any produced water discharges. Suspended solids are frequently present in source waters, particularly surface sources. Since oil and solids occur together in produced water, they will be lumped together in the first part of this discussion.

A question often asked in oil field waterflood asset management is why remove oil and solids prior to reinjecting the produced water? The answer is quite straightforward: we remove oil and solids to maintain well injectivity. Without effective oil and solids removal, injectivity would decline as the formation face and near wellbore formation filter out the oil and solids. This filtration by the reservoir results in blocked pore throats within the reservoir rock which leads to a reduction in formation permeability, and hence, increases the difficulty of getting water into the formation, i.e., the well loses injectivity.

The quality of water required for injection into a subsurface formation will vary from location to location. An acceptable quality is usually defined by the amount of permeability decline which is acceptable while injecting the water. In other words, what is an acceptable loss in injectivity and how fast does that acceptable loss occur? Identifying this parameter will determine the type and degree of treatment required. Loss in injectivity is usually attributed to solids although the influence of oil can be significant.

Suspended solids. This item refers to all suspended solids in injection water, whether the injection water is produced water or sources water. Solids injected into a formation usually reduce the effective pore throat size in the reservoir rock, in essence, causing a restriction or barrier to flow. The solids injected into an injection well can cause a filter cake to be built up on the formation face or can enter the rock pore space and cause restrictions in the pores. Whatever the scenario, the outcome is the same: a loss in effective permeability of the injection zone which usually manifests itself as an increase in injection pressure. To determine the potential injectivity loss caused by solids in the injection water, on-site corefloods are usually carried out. This is the preferred approach to determining the effect of solids but if no suitable core material is available, some "rules of thumb" exist for relating the size of suspended solids to the tendency of the water to plug the formation. These are known as the 1/3-1/7 Rules.

Briefly stated, these rules are as follows:

1. Particles having a median diameter at least one-third the median pore throat diameter will not invade the formation but will bridge and form an external filter cake

2. Particles having a median diameter between one-third and one-seventh the median pore throat diameter can invade the formation causing an internal filter cake which will cause plugging and permeability decline

3. Particles having a median diameter smaller than one-seventh the median pore throat diameter will pass through the porous medium

Use of these rules of thumb to set water quality requirements for injection or a filtration particle size can result in at least two different design criteria. The conservative approach is to set the filter specification so that the mean particle size in the filter effluent is less than one-seventh of the mean pore size. This assumes that neither an external nor internal filter cake is formed. The other approach is to set the filter specification above one-third of the median pore throat size. This allows the formation of an external filter cake, as opposed to an internal cake, which can then be removed by either back flowing the well or treating with acid. The risk of the first approach is that the filtration level is impractical technically and economically. In the second approach the risk is that the build up of an external cake may be so fast as to cause an excessive rate of injectivity decline, the filter cake cannot be effectively removed, or the number of remedial treatments is limited. If we assume that the rules of thumb are valid for a particular injection system, the choice of which approach to follow requires a strong experience factor, and economic considerations. A more acceptable alternative in many operations is to run coreflooding tests to determine filtration requirements.

Dispersed Oil

The effect of oil on permeability decline or injectivity loss is not well defined and may well vary with location and crude type. Possible injectivity impairment can be caused by formation of an emulsion at the formation face, relative permeability effects, or the possibility that oil droplets act like solids and effectively reduce pore throat size in the manner previously described. In many systems the requirements for oil concentrations in the injection water are set by the capabilities of the process facilities and not by measuring the effect of oil in water on formation rock permeability.

In a typical coreflood scenario, oil is removed prior to determining the filtration specification for solids removal. In many systems the filtration specification is set based on coreflood results and the oil concentration is determined by the performance of the filtration system. Many filter systems have a limit to the amount of oil they can tolerate. Most filters are poor oil removal devices and usually have an upper limit of 15 ppm of oil in their influent. Typical filter outlet oil concentrations are below 5 ppm which in many cases may be acceptable for injection.

Monitoring is an important part of a water treatment system. Having determined the required water quality, and designed and constructed a treating system, it is then necessary to determine that the plant is producing the desired water. Monitoring answers such questions as:

"How is the treatment plant operation?"
"Is it operating to the design parameters?"
"Is the water quality adequate for the injection system?"
"What is the effect of operational changes upon system performance?"
"What operational changes could be made to improve efficiency and cost-effectiveness?"

This section is concerned with monitoring of water quality with regard to dispersed oil and suspended solids. Monitoring techniques for corrosion, scale and bacteria are discussed in latter sections as are general principles of chemical analysis that are useful to monitoring.

Corrosion Control

Corrosion control is an important aspect of the operation of waterflood equipment, starting at the producing well or other water source and carrying through all the way to the injector. We will cover some of the major effects of corrosion, and discuss the types of corrosion most often seen in waterfloods. Then we will discuss the rate of corrosion and the effect of the water's composition on this rate.

The most obvious effect of corrosion on water flood equipment is that of limiting the useful life of the equipment. Corrosion rates commonly found in oil field fluids can limit the life to a few years, or even less. A less obvious, but just as important, effect is that on injection water quality. From corrosion due to produced water in wellbores and flowlines, through corrosion in water treating equipment, to corrosion in injection flowlines and wells, each step in the production and injection process must be considered. The presence of dissolved iron resulting from corrosion is one of the major sources of water quality problems. Iron oxides and iron sulfides create problems in oil-water separation, in water filtration and treating, and in water injection. Corrosion products can increase the friction along pipe walls and decrease carrying capacity, and may also contribute to plugging of injection wells. Iron solids interfere with water treatment—as adsorption sites for filming corrosion inhibitors, high levels of these solids increase the concentration of inhibitor needed to protect piping, tanks, and injection wells.

Iron sulfides are preferentially oil-wet. When oil-wet, they have densities near that of produced water, discouraging gravity separation. Oil-wet iron sulfides stabilize emulsions. They are compressible, forming low permeability filter cakes in filtration equipment and injection wells, and they are very difficult to break or dissolve.

Corrosion can be classified into two types, general corrosion and localized corrosion. General corrosion occurs over a large area, resulting in gradual thinning. This is the type of corrosion that is addressed by adding a corrosion allowance to the thickness of equipment walls, so it is rarely a threat to equipment life. However, general corrosion can contribute large amounts of corrosion product to the system.

Corrosion rates of the more noble metals (nickel, chromium, and copper alloys) are generally low enough that dissolved metal ions are not a water quality concern. Also, dissolved iron in itself is not a problem, except in boiler feed water for steam floods. The problem arises when the water contains dissolved oxygen or sulfides, so that iron compounds are precipitated. The most common iron compounds are the oxides (magnetite and hematite) and the many varieties of sulfides.

Several techniques are used to control corrosion in water flood operations. The most important technique is the one that occurs before the facilities are even built, with the proper choice of processes and materials of construction. Corrosion monitoring methods and equipment should be considered early in the design. Of equal importance are construction methods, such as welded versus bolted or flanged, and coating specifications.

Internal protection of tanks and vessels is normally accomplished by a combination of materials selection, coatings, cathodic protection, and chemical treatment (including corrosion inhibition and oxygen removal). Internal protection of pipelines is by materials selection, coatings and linings, chemical treatment, and scraping. External protection above grade is by paint or coatings, and below grade by coatings and cathodic protection.

Corrosive gases normally found in oil field waters include oxygen, hydrogen sulfide, and carbon dioxide. Produced waters normally contain no oxygen, but often have dissolved hydrogen sulfide and carbon dioxide. Surface processing facilities often allow produced water to contact air and pick up oxygen. Natural surface waters and sea water used to supplement produced water are normally saturated with oxygen, but contain little hydrogen sulfide or carbon dioxide.

The simplest way to deal with oxygen in produced water is to exclude it. All vessels in the production system should be closed and, if appropriate, gas blanketed. If the produced water becomes contaminated with oxygen, or if we are using surface or sea water, oxygen scavengers can be used. Three classes of scavengers are common: ammonium sulfite or bisulfite, sodium sulfite or bisulfite, and sulfur dioxide. All three classes work by reacting sulfite/bisulfite with oxygen to produce sulfate ion. This reaction is slow, so that catalysis is sometimes necessary, and allowing some reaction time is helpful. The produced sulfate ion is often converted by sulfate reducing bacteria to sulfide ion, so care must be taken in applying these chemicals.

For large volumes of water saturated with oxygen, it may be more economical to use mechanical means to remove oxygen, such as vacuum deaeration, gas stripping, or steam stripping. Acid gases can be removed with the same methods used to remove oxygen, except that it is rarely possible to exclude the acid gases. Nor can scavengers be used to remove carbon dioxide. There are three types of hydrogen sulfide scavengers: iron sponge, aldehydes, and strong oxidizers. Iron sponge is used mainly to remove very low levels of hydrogen sulfide. The aldehydes are effective, but can be dangerous to use without proper training. The strong oxidizers include chlorine dioxide (as a gas or as a stabilized liquid) and hydrogen peroxide, and can also be dangerous. These chemicals also act as biocides, providing further benefit.

Corrosion inhibitors useful in water flood operations are almost exclusively high molecular weight nitrogen containing compounds known collectively as filming amines. They may be either water soluble or water dispersible, but generally the better choice is a water dispersible chemical, which has better film persistency than the water soluble materials. Corrosion inhibitors are effective at low concentrations in clean water. Generally speaking, we should be able to get adequate protection with 5-10 ppm of inhibitor. Because the volume of water to be treated in a waterflood is usually quite large, there is an economic incentive not to overtreat. In addition, corrosion inhibitors usually contain surfactants that can impact other processes in the injection water system and so their use should be minimized. If the water contains significant amounts of suspended solids, especially those containing iron sulfides, the minimum concentration of corrosion inhibitor necessary to provide protection increases to as much as 25 ppm. This is a good example of the interrelations among the processes required in a waterflood.

Materials of Construction

From the point of view of installed cost, ease of installation, and temperature and process requirements, the construction material of choice is usually plain carbon steel. Coatings, cathodic protection, and chemical treatment can be used to control corrosion, but in some cases it is more economical to use a more corrosion resistant material.

Adding a minimum of 11% chromium to iron results in stainless steel, which spontaneously forms a stable, corrosion-resistant passive film in the presence of water. The addition of nickel improves the hardness, ductility, and weldability. Molybdenum is often added to improve resistance to pitting and crevice corrosion, and to improve the strength. Stainless steels typically perform best in oxidizing environments, which somewhat limits their usefulness in waterflood equipment. They also perform poorly in erosive environments. Nickel alloys, including Monel, Inconel, and Hastelloy, are more corrosion resistant than the stainless steels, and usually are resistant to pitting and crevice corrosion. They are also very expensive. Copper alloys are useful in oxygenated seawater environments, as in the intake systems for seawater floods. They are susceptible to attack by hydrogen sulfide and ammonium, and perform poorly in high velocity waters. The brasses (copper-zinc alloys) show more resistance to flow. Aluminum is resistant to corrosion near neutral pH, but is susceptible to pitting in the presence of chlorides. Several aluminum alloys are available with superior resistance to chlorides. The more exotic alloys of nickel, copper, and aluminum are rarely used as the primary material of construction in oil field equipment. They are more often seen in special applications, such as valve internals or mounting hardware.

Coatings are used to provide corrosion protection to steel vessels. Linings are intended as the primary containment material, but use an outer steel vessel for structural strength. Three types of coatings are commonly used in the oil field. Thin film coatings are 10-20 mils thick, generally applied in two or more coats. Most are based on epoxy resins, but some are vinyl or inorganic zinc based. Some newer coatings are based on elastomeric urethanes at 30-60 mils total thickness. The thin film coatings are used to protect against mild corrosion, or in conjunction with cathodic protection, against severe corrosion. Glass flake reinforced coatings are somewhat thicker, either sprayed in two coats to 30-40 mils or troweled in two coats to 60-80 mils thickness. The glass flakes are embedded in either epoxy or polyester resins. These coatings are for mild to severe corrosion, and are more expensive than thin films. They are also more resistant to erosion and abrasion. The laminate reinforced coatings are 80-125 mils thick, built by hand layup of resin and fiberglass mats. There are generally three resin layers and two mat layers, with a surfacing resin layer and a final gel coat. This is the most expensive coating system, useful against severe corrosion. The most important step in the application of any coating is the surface preparation. A solvent wash removes oily residue, then a water wash removes salt deposits. The surface is then abrasively cleaned to the final cleanliness and profile, or roughness. The coating is applied on the same day as the blasting, if possible, within strict temperature and humidity limits as set by the manufacturer. The film thickness is checked after each coat, and the final product is inspected for holidays. Linings are usually much thicker materials than coatings, with at least some strength, and are used mainly for pipelines. They can be used to repair very corroded pipelines, or to prevent corrosion in very corrosive fluids.

Cathodic protection can be used to protect the internal surfaces of any vessel containing water. The protection applies only to internal surfaces below the water line. External surfaces and surfaces in the oil or vapor zones are not protected. The anodes must also be in the water zone. Anodes in the oil or vapor spaces will not discharge current. The external surface of any buried structure may be protected by buried anodes.

Water/Water Interactions – Scale

Precipitates (solid deposits) may form when (1) two incompatible waters are mixed or (2) when a single water undergoes chemical and physical changes that reduce the solubility of one or more components. In the first case, the two waters each contain one or more ions that when mixed together form an insoluble compound. A common example is the formation of insoluble barium sulfate when an injection water high in sulfate (e.g., seawater, is mixed with a formation water that is high in barium). The second type is illustrated by formation of insoluble calcium carbonate when produced water is depressurized. If the precipitated material adheres to a surface, it is called scale. Precipitated material that does not adhere to surfaces is called suspended solids. In waterfloods where the source water is different from the produced water, precipitates can be formed from mixing of two incompatible waters (e.g., seawater with high sulfate mixing with formation water with high barium). The problem usually occurs in the producing wells or the surface production facilities where mixing occurs. Not much mixing of the waters occurs in the reservoir. Another precipitate found in many oilfield operations is calcium carbonate. This precipitate is formed when produced water loses carbon dioxide when it is brought to the surface, separated from crude oil and treated for discharge or injection. The calcium carbonate appears as both scale deposits and as suspended solids.

Many production problems occur from scale formation. Thus, it is important to keep the waters free from generating solids. Injection water that becomes unstable (e.g., from an increase in temperature as it goes downhole) will generate solids that may adhere to the injection equipment, or worse, plug the formation at the injection well perforations. If the injection and connate brines are incompatible, then solids may deposit where these two waters mix. It is possible for the scale to deposit very near the injection wellbore and hamper injectivity. Also, it is common for the incompatibility to cause problems in the production end of the system. This is because the mixing of the two brines is more intense near the production well after the injection water has traveled a long distance through a heterogeneous reservoir. Brines from different production zones may mix together near the producing well or in the production tubing. Scale deposition may occur if the brines are incompatible; sometimes a change in the environment alone (pressure, temperature, etc.) will be enough to initiate precipitation.

Microbiological Effects

Bacterially related problems. Bacteria play a part, either directly or indirectly, in many of the problems encountered in production operations. When bacteria grow and multiply, they produce large amounts of extracellular polymer that is called biofilm. This polymer has many beneficial effects for the bacteria (1) concentrating mechanism to remove nutrients from the water, (2) storage of nutrients for future use, and (3) protection from adverse environmental factors (biocides). The biofilm contains a large amount of water (usually more than 80%) and has a low permeability. It acts as a glue for other particulate matter. By fouling surfaces and plugging porous media (filters, ion exchange beds, formation rock), the biofilm can restrict flow and even lead to loss of crude oil reserves.

Corrosion rates are frequently greater when bacteria and biofilms are present. The problem is compounded because microbiologically influenced corrosion (MIC) is inevitably localized (pitting) corrosion. Equipment failure from penetration sometimes occurs within a very short time period.

There are numerous examples of production of sufficient hydrogen sulfide by bacteria to turn a reservoir from sweet to sour production. This usually occurs when sulfate reducing bacteria (SRB) are introduced into the producing formation and the other necessary conditions for their growth are satisfied. SRB are capable of converting inorganic sulfate ions into hydrogen sulfide using a variety of organic compounds as the reducing agents. Although the amount of hydrogen sulfide produced by an individual bacteria is low, the number of SRB in a single milliliter of water or on a square centimeter of surface may be in the millions or even billions. Microbially produced hydrogen sulfide concentrations may reach hundreds of parts per million in the gas phase of the produced fluids.

A three step approach can be used to control problems that are caused by bacteria. The first step is to *Culture* to determine (1) are bacteria present that could be causing the problems; (2) what type of bacteria are present; (3) where are they located; and (4) how many are present? The next step is to *Clean*. Bacteria are usually protected by biofilms and other organic and inorganic deposits. For a control procedure to be effective, as much of these protective layers as possible must be removed before application of a biocide. Chemical treatments (acids, oxidants, solvents) or mechanical procedures (brushing, scrapping, pigging) can be used for cleaning. Frequently, combinations are more effective. The third step is to Kill the residual bacteria. Biocides are commonly used for this step. Other procedures that have been used are ultraviolet radiation or even heat sterilization.

A variety of chemicals, organic and inorganic, have been used or proposed as biocides. The most widely used are the oxidizing biocides, with chlorine (or hypochlorite) the most common. Oxidizing biocides are nonselective and usually effective at relatively low concentrations. Their oxidizing power means that they are effective in penetrating and breaking up biofilms. In addition to their biocide action, they are useful for cleaning the system. By the same token, they are consumed by oxidizable material in the system including sulfides. Usually, the oxidant demand must be met before there is sufficient biocide residual to be effective. The inorganic biocides are no longer used to any great extent because of severe environmental consequences. Aldehydes are widely used as biocides. Formaldehyde is losing favor because of its carcinogenic properties. Acrolein is very effective as a biocide, chemical cleaner, and hydrogen sulfide scavenger, but it is very difficult to administer. Aldehydes are reactive with and consumed by sulfides. Systems with high sulfide concentrations would have high chemical consumption. A wide variety of organics and combinations of organics are used as biocides. A new biocide based upon a phosphonium sulfate formulation has been used successfully against sulfate reducers in the North Sea. This chemical is also available in Australia. The supplier is getting EPA approval now for use in the United States.

Water Injection and Production Control

Prior to waterflooding, the total injection rate must be determined for each pattern of the waterflood. This then becomes the desired injection rate to be maintained for that particular pattern. In the case of a multiple-zone waterflood, the desired rate may also be determined for each of the zones within an injection well.

After a waterflood has commenced, the desired injection rate must be monitored. As conditions change within the waterflood, it may be necessary to adjust the injection rate from time to time.

Injection control, discussed in this section, refers to methods used to maintain the desired injection rate. Injection control may apply to the overall injection rate into a well or to the individual zones within a well. Three methods of fluid control are presented in this section:

1. Subsurface fluid control (general methods)
2. Subsurface injection control by mechanical devices, and
3. Surface injection control

The control of high rates of water injection into multiple zones poses a challenging problem. It has been performed successfully in the United States, but at low rates. The control of high injection rates is an emerging technical area as more high rate waterfloods are being installed throughout the world.

In multiple zone waterfloods, it is not only important to maintain the desired injection rate into the overall well, but also to maintain the desired injection rate into the individual zones. Uncontrolled injection may preferentially enter one major zone within a formation and may cause a channel or "thief" to develop in that zone. The channel could then cause premature water breakthrough to offsetting producing wells and may thereby decrease the sweep efficiency in that particular zone. Repeated cycling of water injection into a "thief zone" will cause offsetting producers to become uneconomical, not having swept the other zones in that pattern. Also, once a channel is created through a zone, it is difficult to repair the damage.

Figure 10-3 shows a channel between an injection well and a producing well caused by lack of injection control. Note that the upper Zone A has the highest permeability of 1000 md and the lowest bottomhole pressure of 300 psi. Zone A is shown to be accepting the majority of injected water, is completely flooded out, and is shown to be cycling the injected water to the offsetting producing well. Zone B has a more moderate permeability of 200 md and a higher bottomhole pres-

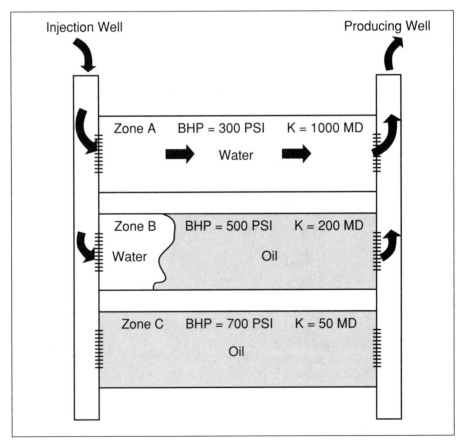

Figure 10-3 Waterflood Case Without Injection Control

sure (500 psi). Consequently, Zone B is accepting smaller amounts of the injected water and produces a smaller amount of fluid in the offset well. Zone C with the lowest permeability (50 md) and highest bottom-hole pressure (700 psi) accepts none of the injected water, and it therefore is not producing any oil in the offset well.

Controlled or regulated water injection can create a much more uniform flood front and oil bank, and can result in a much better sweep efficiency. Injection control will allow all zones in a formation to be processed simultaneously rather than individually.

Subsurface and Surface Fluid Control

Methods for subsurface fluid control involve modifying production or injection profiles of wells to:

- Divert injected fluid from thief zones into previously under-swept zones to improve flood conformance
- Shut off undesirable produced fluids
- Control water or gas coning
- Repair casing leaks
- Repair cement failures

There are two benefits from using profile modification technology in field operations:

1. Increased profitability from increased oil recovery, and
2. Reduced volume and cost of water lifting and disposal in production or waterflood operations

Proper well and process selection, job design, job performance, and placement are important factors for successful results. These items are discussed in more detail in the following sections.

Injection and Production Well Problems

The purpose of this section is to provide information on available processes for injection and production profile control and their applications. Various mechanical, cement, and chemical methods for controlling water injection and production are discussed.

Modification of the production or injection flow profile of a well is desirable in cases of excess water or gas production, as well as in underinvaded zones of secondary or tertiary projects. Figure 10-6 illustrates the types of profile control problems commonly encountered that result in high water production rates or poor waterflood sweep efficiency. These are:

Casing leak. Corrosion holes, collar failures, split casing, which leaks water from an aquifer into the well. This is most often an old-casing problem.

Mechanical failures. A packer leaks water from above or below into the producing zone. Casing shoe leaks.

Poor injection profiles. Uneven water injection caused by permeability variations, fractures, faulting.

Perforations. Plugged, too few, too shallow, or placed out of zone.

Water and gas coning. Water from below or gas from above encroach into the oil production perforations, the result of production drawdown and vertical permeability or fractures.

Natural encroachment. Natural or assisted recovery water drive breaks through the higher permeability zones, increasing the water-oil ratio. This is a natural consequence of oil production.

Cement failure. Provides channels behind casing for extraneous water to enter the production intervals.

All of these problems are targets addressed by the technology that we are about to discuss.

Available Treatments

Currently, cement and mechanical systems are the most widely used methods for permanently controlling water flow. However, cement and mechanical methods are effective only in the wellbore area. In reservoirs with good vertical permeability and no intervening shales, these profile control methods are not very successful. Chemical matrix treatments are used to a lesser extent than cement or mechanical methods, but they can change permeability far out into the formation, so they do not have the shortcomings of the cement and mechanical systems.

In matrix treatment operations, chemical agents are applied as low-viscosity solutions to fill pores of the permeable formation radiating from the wellbore. After a certain time has elapsed, the solution reacts, forming an impermeable or mobility-modified barrier. A mobility-modified barrier allows the permeability to oil in a porous medium to go unchanged, while reducing the permeability to water.

Permanent Plugging, Injection or Profile Modification Systems[4-24]

We will consider six general systems for permanently plugging or modifying the wellbore, the formation matrix, or both. These systems and their plugging properties are listed in Table 10-3.

Table 10-3 Permanent Agents-the Utility of a Specific Agent Is Dependent on its Plugging or Mobility Modifying Characteristics

Type of Mechanical/ Chemical Treatment	Wellbore Plugging Agent	Matrix Plugging Agent	Matrix Mobility- Modifying Agent
1. Mechanical	x		
2. Cement Squeeze and Plugback	x		
3. Sodium Silicate Gels		x	
4. Non-Crosslinked Polymer Solutions			x
5. Crosslinked Polymer Gels		x	
6. Solid Thermosetting Polymers	x	x	

1. Mechanical Methods
One of the most positive ways to repair casing leaks or to exclude unwanted perforations is to run an inner liner, cement it in place, and reperforate zones of interest. This is especially useful for badly corroded casing. One must keep in mind that an inner liner will have no effect outside the old casing, so if the problems being treated stem from poor casing cement, channels, fractures, vugs, etc. that contribute to vertical fluid movement outside the wellbore, an inner liner will do nothing but exclude fluid entry until perforated.

A casing patch can be used to exclude water entry over short casing intervals (practically, less than 70 feet). The casing patch is a length of corrugated pipe with an epoxy-painted fiberglass sheath around it. The casing patch is run on tubing to the interval to be repaired, holddown buttons are activated, and an expanding cone is hydraulically pulled through the collapsed patch to swage it into place. The epoxy-fiberglass sheath extrudes through the holes and

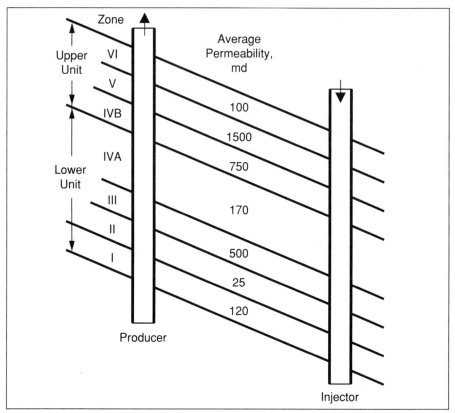

Figure 10-4 Field Exhibits Water Control Problems Due to Permeability Variations

sets up to form a seal between the patch and the casing. These patches are used primarily to repair corrosion holes, collar failures, and split casing.

Many reservoirs are made up of discrete oil zones, separated by substantial shale barriers, as shown in Figures 10-3 and 10-4. In this particular case, the different oil zones have substantially different permeabilities, resulting in early waterflood breakthrough, excessive water production, and poor waterflood conformance because of thief zones in the water injectors. An effective method for reducing water production from this reservoir is the use of selective equipment completions. An example is shown in Figure 10-5. Figure 10-6 shows different types of well problems that may be present in a waterflood.

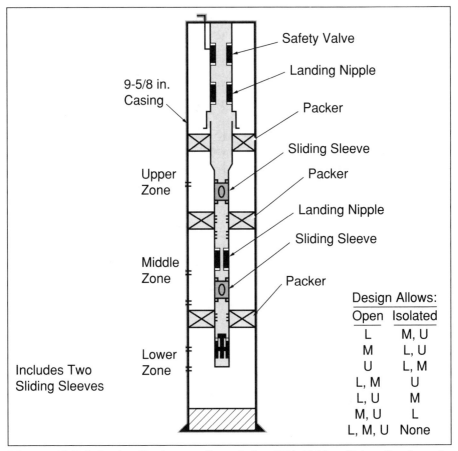

Figure 10-5 Selective Equipment Completion With Tubing String Continued Through the Well

Through the use of wireline plugs and sliding sleeves, the three producing zones can be opened to or isolated from production in any combination. For the injection wells, replacing the sliding sleeves and plug with side-pocket mandrels with wireline-retrievable flow control valves provides a method to control injection profile and conformance.

It is worth repeating that these mechanical methods and the cementing methods discussed in the next section affect only fluid movement into and within the wellbore. Casing cement channels, vertical permeability, fractures, and vugs can all defeat the control from these in-hole devices, as shown in Figure 10-6.

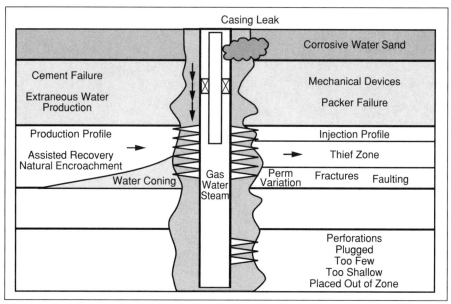

Figure 10-6 Types of Well Problems

2. Cement Methods

Cement squeezes and plugbacks are noted above. These techniques are commonly used for plugging perforations and abandoning lower zones in a well to restrict water flow. Advantages of cement are its strength and our familiarity with cement. Cement's greatest disadvantage is that it cannot penetrate the formation matrix, so its effect is limited to within the drilled hole. A cement squeeze will generally require a rig to pull all equipment out of the hole, run a drillable bridge plug, run tubing with a cement retainer, squeeze, drill out, then rerun production equipment. Squeeze and plugback treatments are effective only if water-producing or thieving intervals can be identified and isolated, and if the formation prevents water from bypassing the treated interval. Figure 10-7 illustrates water bypassing a cement plugback.

Diesel oil cement and slurry oil squeeze are water shutoff systems. Diesel oil or kerosene and dry cement are slurried at the surface and pumped downhole to fill channels, fractures, or perforations. On contact with water, the cement hydrates and hardens, sealing the openings. These systems must find water to work, so if the formation is dry or if too much oil preflush is injected, the system will not form a useful plug.

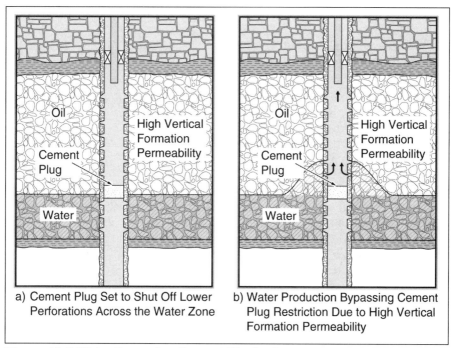

a) Cement Plug Set to Shut Off Lower Perforations Across the Water Zone

b) Water Production Bypassing Cement Plug Restriction Due to High Vertical Formation Permeability

Figure 10-7 Water Bypassing Cement Plugback

Some service companies market Superfine Cement, an extra-fine-grind cement, which as a slurry, can penetrate further into a coarse matrix than can regular cement. This product is being used in California for plugging the gravel-filled annulus in gravel packed wells. A diesel oil cement system using Superfine Cement is being developed.

3. Sodium Silicate Gels

Profile modification gels contain as much as 99+% water, which is immobilized in a network of organic or inorganic polymers. Sodium silicate solutions in formation matrix can be catalyzed by acids or bases to form a jelly-like mass that prevents fluid flow through the formation. The extent and rate of gelation are a function of SiO_2 concentration, pH, salinity, and temperature. The sodium silicate treatment solution usually has a room temperature viscosity of about 15 cps.

Sodium silicate gels have been successfully used to prevent water coning from an underlying active aquifer, and to correct vertical conformance in injection and production wells. They can be applied over a wide range of temperatures, pressures, and salinity. Their low cost allows larger volume treatments. The low-viscosity aqueous system tends to go where the water is. Disadvantages are that sodium silicate gels are not very strong and dissolve slowly if water shutoff is not complete. For these reasons, sodium silicate treatments are often backed up with a cement squeeze to form a strong plug at the perforations, which isolates the wellbore from the treated matrix.

4. Polymer Systems
Use of polymeric systems for profile modification or plugging is increasing.[24] When a production well is treated, it is usually to reduce water production and water-oil ratio (WOR). Such treatments are often limited to the near-wellbore vicinity. When an injection well is treated, it is usually to improve the volumetric sweep efficiency. In this case, a permanent, in-depth change in reservoir permeability is desired.

There are two major types of polymer systems used for profile modification and plugging:

- Non-crosslinked polymer solutions
- Crosslinked polymer gels

5. Non-Crosslinked Polymer Solutions
Polyacrylamide solution (2,000 to 4,000 ppm in water) reportedly acts as a mobility modifying agent, reducing formation permeability to water while having little effect on the permeability to oil. Treatments typically extend 10 to 30 feet from the wellbore. Knowing where the water is entering the well and treatment zone isolation is not as important as with other processes.

Several mechanisms are thought to cause the reduction of water production by the polymer solution. First, the polymer solution should go predominantly into the zones with high water saturation. The polyacrylamide may sorb from solution to form a hydrophilic film on the matrix, as shown in Figure 10-8. Water passing near this film is slowed by attraction to the polymer.

6. Crosslinked Polymer Gels
Crosslinking polymer solutions to form a gel provides better physical plugging of pores. This is achieved with polyacrylamide solutions by adding multivalent ions that form a three-dimensional network, which entraps the solution's water.

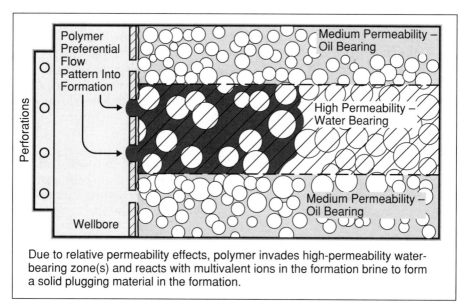

Due to relative permeability effects, polymer invades high-permeability water-bearing zone(s) and reacts with multivalent ions in the formation brine to form a solid plugging material in the formation.

Figure 10-8 Physical Plugging Theory

Two polyacrylamide gel systems are in common use. Both are based on the controlled release of multivalent metal ions, which react with a polymer to form a gel. The first process, the Chromium Redox Process, consists of injecting a polymer slug containing chromium (VI) cations and a chemical reducing agent. The reducing agent reduces the chromium (VI) to chromium (III). Chromium (III) reacts with the polyacrylamide to form the gel. Solution gel (and pump) time depends upon the rate of chromium (VI) reduction. Concentration and the type of reducing agent can be varied to control gel times from a few minutes to several days.

The second system, a combination process, involves the sequential injection of polymer solution, aluminum citrate solution, then a polymer solution, followed by resumption of water injection. Permeability reduction by this process is thought to be caused by the formation of a layered polymer/aluminum ion gel network. Polymer adsorbed on the matrix from the first polymer slug forms the base layer. Aluminum ions attach to this base layer from the aluminum citrate solution. A second polymer layer forms through interaction between retained aluminum ions and the second polymer slug as it flows by. An advantage of this process is the large permeability reductions obtained using dilute solutions.

Adding crosslinking agents to polyacrylamides generally produce physical plugs, which resist flow of all fluids through formation pores. This results in a more permanent plugging effect but may restrict flow of oil or gas as well as water. It is generally recommended that crosslinked polyacrylamides be used as grouting materials and be injected only into zones that need to be plugged or restricted, such as thief zones in injection wells and watered-out zones in production wells.

Some operators also utilize solid thermosetting polymers, where polymers set as physically impermeable and solid plugs as a result of higher temperature encountered in the reservoir.

Treatment Process Selection

Selection of a profile modification process depends on the problems being treated and the desired result.[25] It is most important to determine:

1. What the well problem is
2. What is causing it
3. Where the well should be treated to correct the problem

Determining the cause of the problem and its location is most often not simple. The production logging section of this course will present a number of ways to get a better definition of the problem. By identifying the problem (i.e., water coning, steam thief zone, casing leak), the desired result of the treatment is also identified.

Two types of problems are usually encountered:

1. Conditions beyond the wellbore such as matrix permeability variation, fractures, and coning, and
2. Wellbore completion factors such as primary cement, casing, and mechanical devices

Table 10-3, 10-4 and 10-5 summarizes the types of processes used to remedy such problems.

In profile modification, there are two major questions:

1. Over how much of the open interval and to what depth into the formation does the permeability need to be modified to achieve the desired results?

2. What can we achieve technically, practically, and economically?

Table 10-4 Remedial Treatments

| Problem | Cement Plug or Squeeze Treatments | Chemical Matrix Treatments | | | |
| | | Plugging Agents* | | | Mobility Modifier** |
		Sodium Silicate	Solid Polymer x	Cross-linked Polymer Gels	Non-Cross-linked Polymer Solution
I. Conditions beyond the wellbore (i.e., reservoir formation/ rock matrix) A. Producing wells (shut off of undesirable fluid flow from coning, natural influx, or assisted recovery	Note: Cement plug or squeeze, and mechanical plugging are only effective for Group I problems if the formation prevents the injected or produced fluid from bypassing the treated interval				
Water	x	x	x		x
Gas	x	x	x		
B. Injection wells (diversion of injected fluid into previously underswept zones)					
Water thief zone	x	x		x	x
Steam thief zone	x	x	x		
II. Wellbore completion factors					
Cement failure channelling	x		x		
Casing/tubing leaks	x		x		
Mechanical device failure					

x = Applicable
* = Requires isolation placement technique to protect the oil-bearing zone in producing wells
**= Selective reduction of the formation matrix relative permeability to water, with the permeability to oil unchanged

Table 10-5 Well Treatment Selection Information

Formation
Lithology | Temp °F Reservoir
Average Depth, Ft | Bottomhole
Productive Acres | Permeability Horiz Max
Initial Oil in Place | Min
% Recovery Primary | Ave
Secondary | Vertical
Oil Gravity, API | Dykstra-Parsons Perm Var
Viscosity in Res Oil | Porosity in Thief Zone
Water | Residual Oil Saturation
Drive Mechanism | Water Saturation
Current Reservoir Pressure | Clay Types and Percent
Current Water Oil Contact | Fractures?
Current Gas Oil Contact | High Perm Streaks?

Water
Connate Water Analysis
Source of Injection Water
Injection Water Analysis
Current Treating Techniques

Wells — General
Completion Date | Net Oil Thickness, Ft
Completion Type | Thief Zone Description
Perforated Intervals | Treatment Intervals
Bottomhole Temperature | Latest Acid Treatment
Sand Thickness, Ft

Injection Wells | **Production Wells**
Wellhead Injection Pressure | Source of Water Breakthrough
Bottomhole Pressure | Initial Oil Water Prod Rate
Average Injection Rate | Current Oil Water Prod Rate
Date of Last Profile Log | Cumulative Oil Production
 | Type of Lifting System
 | Fluid Level in Wellbore
 | Treatment Objectives
 | Post Treatment Oil Rate
 | Post Treatment Water Rate

Near-wellbore treatments in injection wells are effective for diverting water in the near wellbore area, but further into the formation, water may find the path of least resistance back to the high-permeability zones or channels. Thus, only a very small part of the reservoir may be affected by a near-wellbore treatment. To achieve the desired results, it may be necessary to inject treatment fluids into the formation as far as possible. This will increase job costs, but it also increases chances for success.

Any formation damage in a production or injection well should be removed prior to a matrix treatment. If scale, asphaltine, paraffin, or sand are partially plugging the well during a treatment, chances of success are reduced, and results will almost always be disappointing. If using a work string or coiled tubing, it too should be free of debris. If a problem is suspected, it may be wise to acidize, wash the well, or reperforate to remove formation damage. When this is done, it may be difficult to determine whether an increase in oil or gas production is due to clean-up operations or to the treatment, but it is the end result (oil in the tank) that counts.

Make-up water for any matrix treatment must be clean, always filtered, and compatible with the formation and the treatment fluids. During the job, the mix water should be monitored continuously to be certain that solids are being removed by the filters. It seems incongruous that the absence of particulates is so important when the intent is to reduce formation permeability. However, if solids are injected with the polymer solution they may prevent a complete treatment.

Selection of Wells to Be Treated

It is critical that the candidate wells have the potential to produce hydrocarbons. For example, in a waterflood if water injection prior to treatment is near the formation fracture pressure, then the rate of oil production may not be improved by improving the injection profile.[27] If the pressure drop between the injection and production wells is not increased, flow through the tighter zones remains the same regardless of flow reduction in the thief zone. Benefits in this case would be limited to savings from smaller water volumes and water lifting and disposal costs. Conversely, if the thief zones take water at an injection pressure that is much lower than the fracturing pressure, it is possible to increase the rate of production by plugging or reducing the permeability of the their zones. Then a higher pressure drop at the same injection rate may result in increased injection into the lower permeability zones and an increase in oil production.

Water shut-off candidate wells. When selecting production wells for water reduction treatments, not all high water cut wells are candidates. To predict the likelihood of a successful water control treatment, analyses of a well's current production and reservoir characteristics are necessary. The following reservoir properties or conditions are useful guidelines[28] for selecting the best candidates for produced water treatments:

1. Production wells are best
2. A high WOR (particularly where this high ratio results from water coning or from a high-permeability thief zone)
3. Water-drive reservoirs, which initially had high oil production and little or no water production, and still have good remaining recoverable reserves
4. A definite oil-water contact (can be estimated with a pulsed neutron survey)
5. Water production from matrix flow, or at least controlled by matrix flow
6. Heterogeneous formations with water zones more permeable than the oil zones (can be determined by gamma-ray neutron log and/or core data)
7. A high fluid level in the well

Subsurface Injection Control by Mechanical Devices

Subsurface injection control is another method of controlling the injection of water in a waterflood. Subsurface injection control in most cases utilizes multiple packers to isolate individual zones or combinations of zones for injection. Subsurface injection control may also be performed with a single packer and limited entry perforating, as will be discussed later. A mandrel and a subsurface valve or some other type of flow control device are installed between each pair of packers to control the flow of fluids into each isolated interval. A mandrel here is used to refer to the receptacle that houses the subsurface valve. Figure 10-9 shows a typical subsurface injection control installation into three different zones.

As Figure 10-9 shows, water is injected down the tubing, and a portion of the total injection volume exits at each of the injection control devices and enters each of the isolated zones. A valve within each injection control device is used to allocate a portion of the total injection volume to each of the zones. The total injection rate into the well is the sum of all of the injection rates into the individual zones.

In actual use, the downhole equipment can be simple for low injection rates or for low maintenance types of installations. However, the downhole equipment can be fairly complex at very high injection rates or for hostile types of environments. The most difficult problem in subsurface injection control is the need to control the rates of injection into each individual zone independently of the injection rate into all of the other zones. This problem becomes more difficult with the total number of isolated zones to be injected and with higher total injection rates.

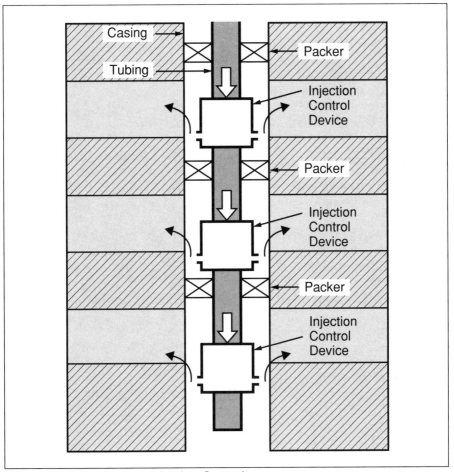

Figure 10-9 Subsurface Injection Control

Subsurface injection control utilizes valves that are similar to the valves used in surface injection control. These valves may be either regulators or orifice valves. The subsurface regulator valve functions just like a regulator, described above in the Surface Injection Control section. The subsurface regulator controls the injection rate independently of pressure fluctuations. The subsurface orifice valve functions like a positive choke, which was described above in the Surface Injection Control section. An orifice valve controls flow by causing a pressure drop through a fixed orifice within the valve body. Under most waterflood conditions, the rate through an orifice valve changes with pressure fluctuations and may, therefore, require frequent changes to maintain a constant rate.

Subsurface injection control is applicable in waterflooding when it is desirable to regulate or control flow into individual zones of a multiple-zone waterflood. Subsurface injection control attempts to prevent the over injection of fluid into each individual zone of a well, as opposed to surface injection control, which is used to prevent the over injection into the well as a whole.

A waterflood, which contains multiple zones with uniform permeability and equal bottomhole pressures, would probably not need subsurface injection control. Injection into the uniform zones would behave similarly to injection into a single zone and could thus be accomplished by surface injection control alone. However, subsurface injection control becomes more of a necessity when multiple zones with high permeability variations and bottomhole pressure variations are waterflooded. Subsurface injection control may also become necessary when surface control methods have failed, allowing thief zones to dominate an injection profile.

Subsurface and surface injection control methods are not utilized together. Although both methods could be used together, they are redundant and will result in additional pressure drops that are not necessary.

The main disadvantages of using any method of subsurface injection control are the high initial installation cost and the high cost of fishing the equipment when it fails. The initial investment depends upon how many zones are to be isolated and the type of selective injection equipment used. An even larger investment may be necessary when the equipment fails, which in some cases can result in a complicated fishing job.

Surface Injection Control

Surface injection control, as the name implies, refers to injection control methods performed above the ground or on the surface. This is in opposition to subsurface injection control that is performed downhole. Surface injection control is usually performed at the surface location of an injection well.

Figure 10-10 shows a typical installation for a surface injection control device. The device is installed in the injection line loop at a wellsite and controls the injection rate before it enters the tubing string. Surface injection control devices serve the purpose of controlling the overall injection rate for a single tubing string. These devices can also be used with dual tubing strings, if a surface device is installed for each string.

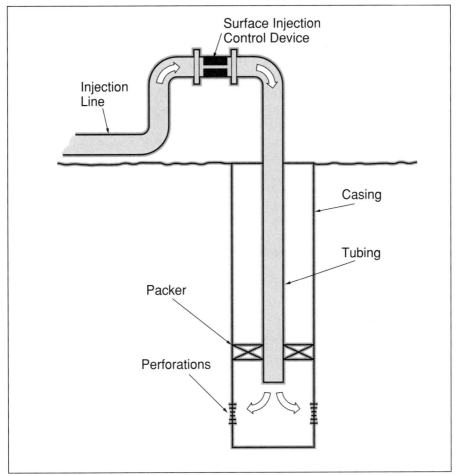

Figure 10-10 Location of Surface Injection Control Devices

The two types of surface injection control devices available for water injection are surface chokes and surface regulator valves. Chokes control water injection by causing a pressure drop through a restriction within the valve. Chokes may require adjustments if the injection system pressure changes. Regulator valves on the other hand, contain a spring-operated piston arrangement, which makes the valve self-adjusting in order to maintain a preset injection rate.

Both surface chokes and surface regulators are installed in the injection line to a well. Both of these types of injection-control devices are relatively inexpensive to purchase and install as compared to sub-surface injection-control devices or dual tubing strings. The investment in surface injection-control devices may only be a small fraction of that for other injection-control methods. Surface injection-control devices also have the advantage over subsurface devices in that they are easily repaired or replaced, which can result in much lower operational costs than other injection-control methods.

Surface injection-control methods, however, may not adequately control water injection if multiple zones are being waterflooded. This is because surface devices control the overall injection rate through a tubing string but cannot control the distribution of the fluid once it reaches the formation. The distribution of injected fluid into individual zones is dependent upon the reservoir characteristics and may, in some cases, require subsurface regulation devices.

Surface methods should always be tried as the first method of injection control. The reason for this is that the investment is small for the installation, and they are very easily removed if they are not successful. Other injection-control methods may require a relatively large initial investment to install, and the cost to remove them can be substantial if they fail.

Thought Items

- In some waterfloods, oil is produced after breakthrough by being dragged along with the produced water or steam. Such wells will be water shutoff candidates only if you can afford to lose this subordinate production.
- If a production well has a uniform, high water saturation over the total producing interval, the well is not a good candidate for a produced water reduction treatment. If you shut off the water, you will most likely also shut off the oil.
- A polymer gel matrix treatment alone won't improve the water injection profile in a well if the water is injected above fracture pressure. Instead, tail in with something strong to plug the perforations and isolate the formation from high pressure. Cement, solid polymers, or a casing patch might do it.
- It takes about a quart of cement slurry to plug a perforation. If you are squeezing more than this, you may be fracturing it away!
- If you displace a fluid down tubing with a denser fluid, the heavier fluid may reach bottom first. Use a wiper plug to separate the fluids or change their densities.

- Plugged perforations will indicate a poor injection profile in the best of injection intervals. Be on guard against this. You may first want to do a stimulation treatment.
- When doing a radioactive injection profile survey, test above and (if possible) below the perforations for an RA indication of fluid flowing up or down behind casing.
- Properly abandoning a watered-out production well may be your best choice for a remedial treatment.
- When you shut in a well, sometimes fluids don't stop moving around downhole. Cross-flow can ruin the best planned remedial treatment.
- If you are injecting all the water you can while staying below fracture pressure, plugging water thief zones will reduce the water injection rate and maybe water production, but may not increase oil production.
- A small amount of solids in your waterflood injection water will give you injection profile leveling right down to zero barrels of water per day.

Conversion versus Newly Drilled Wells

This section deals with the subject of whether it is better to convert existing producing wells or to drill new wells for injection purposes. Sometimes it is necessary to drill replacement wells for injection because the old wells are in poor mechanical condition. The old wells are then plugged and abandoned. However, here new wells refer to new locations that are added to complete waterflood patterns. These wells reduce the spacing that existed between the old wells. The decision then must be made as whether to convert old producers to injectors and to drill new producing wells or to keep the existing producers and drill new injectors.

Many operators consider injectors as secondary in importance to producers. After all, injectors produce no oil directly. Therefore, old producers are usually converted to injectors, and any new wells become producers. In fact, usually the worst producers are converted to injectors. These may include the producers with the highest WORs, lowest production rates, and even dry holes. However, it should be considered that poor producers usually make poor injectors.

The decision whether to convert existing wells to injectors or drill new injectors should be based upon economics. The decision should be part of an overall waterflood plan that yields maximum economic oil recovery.

Several other factors should be considered in making a decision concerning conversion of existing wells versus drilling new wells:

1. Bottomhole location
2. Casing size
3. Casing condition
4. Completion technique

A well's bottomhole location is the first factor to consider when converting existing wells to injection. Many times the surface location is thought to reflect an accurate bottomhole location. However, due to natural drift while drilling and past drilling practices, the bottomhole location may be some distance from the surface location. In this case, the existing well that is being considered for conversion to injection may have a bottomhole location that is out of pattern. An example of this is shown in Figure 10-11.

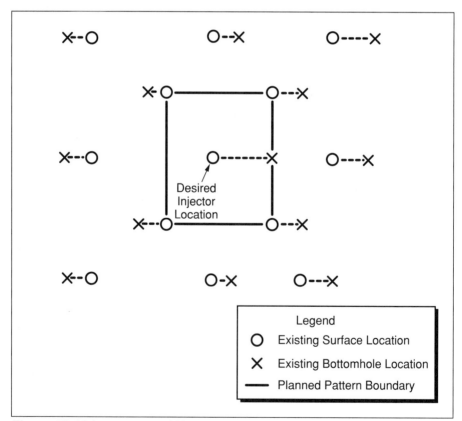

Figure 10-11 Importance of Knowing Bottom-hole Location

Conversion of this well to injection will impact the recovery of that pattern. In this case, a newly drilled injection well could be planned in order to place the bottomhole location in the center of the pattern.

The proper injection pattern should take advantage of the existing well patterns and require a minimum number of new wells. Injection patterns should accommodate known geological features, such as natural fracture trends and directional matrix permeability. New injection wells may need to be drilled in order to take advantage of these geological features and to maximize oil recovery.

Casing size is the next factor to consider in converting wells to injection. The casing size of existing wells should be of sufficient size to allow the desired pattern injection rate. Thus, the casing should be large enough to permit the installation of the proper size tubing string. If dual tubing strings are planned, the casing must be large enough to accommodate them. Also, if subsurface injection control devices are to be installed, such as side-pocket mandrels, the casing should be large enough to permit the proper size of subsurface devices.

Another consideration, as to casing size of existing wells, concerns corrosion. An existing well with small casing may prevent the installation of a liner in the future. A liner may be necessary, if the casing becomes badly corroded from the injection of corrosive fluids. The drilling of new injection wells would permit the proper size and weight of casing to be installed, along with the proper metallurgy for a longer wellbore life.

Casing size considerations for conversion candidates are:

1. Large enough for proper tubing size
2. Large enough for dual strings, if needed
3. Large enough for subsurface injection control devices, if needed
4. Large enough for liner installation in corrosive environments
5. A new well permits proper casing design with respect to:
   Casing size
   Casing weight, and
   Casing metallurgy

Casing condition is the next factor to consider in converting wells to injection. The condition of the casing should be determined in each well being considered for conversion. This may require a pressure test of the casing using tubing and packer, a casing inspection log, or other methods of testing the casing's integrity.

Production wells with worn or corroded casing may not currently pose problems. However, an injection well requires good casing for a proper packer seating. Within the United States, the casing of injection wells must be able to withstand a mechanical integrity test to prove that shallower formations containing potable water are not being polluted.

Existing wells with casing in poor condition may require cement squeezing, casing patches, liner installation, or other methods of casing repair. This may greatly increase the cost of using an existing well.

Completions techniques are the next factors to consider in converting wells to injectors. Techniques that were used to complete the existing wells should be compared with current completion practices. Many improvements in completion techniques have been made in the past 20 to 30 years. If the wells are very old, the primary cement job, perforations, stimulation methods, and many other completion practices have experienced many improvements.

Past cementing techniques should be evaluated for all wells being considered for conversion to injectors. Channels behind pipe due to a faulty primary cement job may cause crossflow between injection intervals or injection into other formations. Channels behind pipe could be expensive to locate and repair. Within the United States, an inadequate amount of surface casing may require squeeze cementing of the production casing in order to protect fresh water formations.

If existing wells were originally completed openhole, they may make very good injection wells as far as rate is concerned. Their injection rate may be greater than comparably cased injectors due to the openhole's greater surface area. However, if a poor injection profile occurs, a liner may need to be installed. In large openholes, a good cement job can be difficult to obtain with a liner.

If the existing wells were cased and perforated, the number, size, and location of the perforations should be evaluated. Too few perforations or too small perforations could require reperforating the injection interval. Too many perforations and too close spacing between perforations can create problems in multiple-zone waterfloods. This can make cement squeeze work difficult if it is needed to improve the injection profile. Also, good packer seats are required for installation of selective equipment if this equipment becomes necessary. Also, perforations in gas caps or other formations may require cement squeezing or isolation with packers.

If an existing well is converted to injection, proper cleaning of the wellbore and stimulation can greatly improve the injection rate and profile. Existing producers may contain scale, corrosion byproducts, oily sludge, and many other substances that could hinder injection. Therefore, the wellbore should be thoroughly cleaned, the formation stimulated, and a clean injection string installed prior to injection. A typical procedure for converting existing wells to injectors is shown below:

1. Remove existing tubing and steam clean if they will be rerun
2. Clean out wellbore to PBTD using bit and casing scraper
3. Break down or circulate and wash perforations
4. Treat perforations with acid/solvent mixture to remove scale and oily deposits
5. Stimulate formation matrix with acid/solvent mixture if needed to remove deeper damage, and
6. Pickle tubing string if bare tubing is used for injection string

Methods Of Increasing Injectivity[26]

The following methods have been utilized for increasing injectivities of wells:

- Backflow by pumping, swabbing, dry DST or nitrogen cushion
- High rate – pressure water injection, especially for wells completed in aquifers (normally used for low permeability zone)
- Acidizing, especially in the presence of carbonate scales
- Chemical washing, using chemicals/surfactants
- Steam injection in zone with residual oil saturation, including some plugging caused by deposition of wax or solid hydrocarbons

References

1. Tinker, G. E., "Design and Operating Factors That Affect Waterflood Performance in Michigan," SPE Technical Paper No. 11134, presented at the SPE Annual Meeting, New Orleans, September 1982, SPE, Richardson, TX.
2. Krumrine, P. H. and S. D. Boyce, "Profile Modification and Water Control With Silicate Gel-Based Systems," SPE 13578, International Symposium on Oil Field and Geothermal Chemistry, Phoenix, Arizona, April 9-11, 1985.

3. Dunlap, D. D., J. L. Boles and R. J. Novotny, "Method for Improving Hydrocarbon Water Ratios in Producing Wells," SPE 14822, Seventh SPE Symposium of Formation Damage Control, Lafayette, Louisiana, February 26-27, 1986.

4. Needham, R. B., C. B. Threlkeld and G. W. Gale, "Control of Water Mobility Using Polymers and Multivalent Cations," SPE 4747, SPE Improved Oil Recovery Symposium, Tulsa, Oklahoma, April 22-24, 1974.

5. Bang, H. W., "A Method of Estimating the Volume of Well Treatments," SPE 12935 - Unsolicited Forum note.

6. Avery, M. R., L. A. Burkholder and M. A. Gruenfelder, "Use of Crosslinked Xanthan Gels in Actual Profile Modification Field Projects," SPE 14114, SPE 1986 International Meeting on Petroleum Engineering, Beijing, China, March 17-20, 1986.

7. Nagra, S. S., J. P. Batycky, R. E. Nieman and J. B. Bodeux, "Stability of Waterflood Diverting Agents at Elevated Temperatures in Reservoir Beds," SPE 15548, SPE 61st Annual Technical Conference, New Orleans, LA, October 5-8, 1986.

8. Wagner, O. R., W. P. Weisrock and C. Patel, "Field Application of Lignosulfonate Gels to Reduce Channeling, South Swan Hills Miscible Unit, Alberta, Canada," SPE 15547, SPE 61st Annual Technical Conference, New Orleans, LA, October 5-8, 1986.

9. Townsend, W. R., Steven A. Becker and Charles W. Smith, "Polymer Use in Calcareous Formations," SPE 6382, SPE 1977 Permian Basin Oil and Gas Recovery Conference, Midland, Texas, March 1~11, 1977.

10. McLaughlin, H. C., John Diller and Hugh J. Ayers, "Treatment of Injection and Producing Wells with Monomer Solution," SPE 5364, SPE Regional Meeting, Oklahoma City, Oklahoma, March 24-25, 1975.

11. Sparlin, D. D. and R. W. Hagen, Jr., "Controlling Operations - Water in Producing Materials," and Part 4 - "Grouting Techniques," *World Oil*, June 1984, pp. 149-154.

12. Knapp, R. H. and M. E. Welboum, "An Acrylic/Epoxy Emulsion Gel System for Formation Plugging: Laboratory Development and Field Testing for Steam Thief Zone Plugging," SPE 7083, SPE Fifth Symposium on Improved Methods of Oil Recovery, Tulsa, Oklahoma, April 16-19, 1978.

13. Hess, P. H., C. O. Clark, C. A. Haskin and T. R. Hull, "Chemical Method for Formation Plugging," *Journal of Petroleum Technology, JPT*, May 1971, pp. 559-564.

14. Anderson, G. W., "New Methods Make Downhole Liquid Plugging Practical," *World Oil*, Feb. 1, 1978, pp. 37-43.

15. Hess, P. H., "One-Step Furfuryl Alcohol Process for Formation Plugging," *Journal of Petroleum Technology*. October 1980, pp. 1834-1842.

16. McAuliffe, C. D., "Oil-in-Water Emulsions Improve Fluid Flow in Porous Media," SPE 3784, SPE Improved Oil Recovery Symposium, Tulsa, Oklahoma, April 16-19, 1972.

17. Fitch, J. P. and Rick B. Menta, "Chemical Diversion of Heat Will Improve Thermal Oil Recovery," SPE 6172, SPE 51st Annual Fall Technical

Conference, New Orleans, Louisiana, Oct. 3-6, 1976.

18. Gall, J. W., "Steam Diversion by Surfactants," SPE 14390, SPE 60th Annual Technical Conference, Las Vegas, Nevada, Sept. 22-25, 1985.

19. Marsden, S. S., "A Review of the Use of Foam-Generating Surfactants for Improved Injection Profiles and Decreased Gravity Override in Steam Injection," Department of Petroleum Engineering, Stanford University.

20. Arshad A. and J. H. Hauser, "Enhanced Volumetric Sweep Efficiency," SPE 14291, SPE 6th Annual Technical Conference, Las Vegas, Nevada, Sept. 22-25, 1985.

21. Friedmann, F. and J. A. Jensen, "Some Parameters Influencing the Formation and Progation of Foam in Porous Media," SPE 15087, SPE 56th California Regional Meeting, Oakland, California, April 2-4, 1986.

22. Millhone, R. S., "Subsurface Fluid Control Survey-1975," COFRC Internal Memorandum.

23. Ford, W. O. Sr. and W. F. N. Keldorf, "Field Results of Short Setting Time Polymer Placement Technique," *Journal of Petroleum Technology*. July 1976, pp. 749-756.

24. White, J. L., Goddard, J. E. and H. M. Phillips, "The Use of Polymers to Control Water Production in Oil Wells," *Journal of Petroleum Technology*. February 1973, pp. 143-150.

25. Wade, J. E.: "Techniques for Treating and Completing Water Injection Wells in California," JPT, January 1967, pp. 49-53.

26. Wade, J. E.: "5 Ways to Boost Injectivity of Waterflood Input Wells," *World Oil*, February 1, 1969, pp. 21-24.

11

Waterflood
Project Economics

Waterflood asset management requires economic evaluation of the project. Making a sound business decision requires that a project will be economically viable, i.e., it will generate profits satisfying the economic yardsticks of the company.

The tasks in project economic analysis (See Figure 11-1) require team efforts consisting of:

1. Setting economic objective based on company's economic criteria. Production and/or reservoir engineers are responsible for developing economic justification with the input from management
2. Formulating scenarios for project development. Engineers and geologists are primary contributors with management guidance
3. Collecting production, operation and economic data
4. Making economic calculations. Engineers and geologists are primarily responsible
5. Making risk analysis and choosing optimum project. Both engineers and geologists are primarily responsible for analysis. Engineers, geologists, operations staff, and management work together to decide on the optimum project

This chapter provides a review of commonly used economic criteria, and a working knowledge of analyzing project economics.[1-6] The five cases considered for the example waterflood design problem presented in chapter 7 were analyzed to determine the potentially most economically viable development plan.

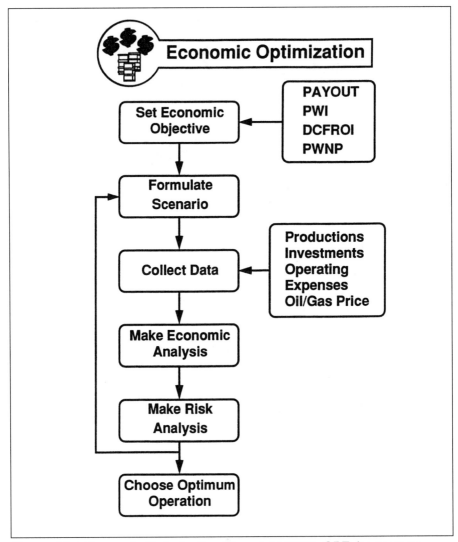

Figure 11-1 Economic Optimization (Copyright 1992, *SPE*, from paper 22350)

Economic Criteria

Making a sound business decision requires yardsticks for measuring the value of proposed investments and financial opportunities. Each company has its own economic criteria with required minimum values to fit its strategy for doing business profitably. Acceptance or rejection of individual proposals are largely governed by the company's economic criteria. Commonly used criteria are reviewed as follows:

Payout Time

The time needed to recover the investment is defined as the payout time. It is the time when the undiscounted or discounted cash flow (CF = revenue - capital investment - operating expenses) is equal to zero.

The shorter the payout time (2 to 5 years), the more attractive the project. Although it is an easy and simple criterion, it does not give the ultimate life-time profitability of a project and should not be used solely for assessing the economic viability of a project.

Discounted cash flow means that a deferment or discount factor is used to account for the time value of money by converting the future value or worth of money to the present worth (PW) in accordance to the specified discount rate. The time value of money is not recognized in case of undiscounted cash flow.

Considering that revenues are received once a year at the midpoint of the year, the discount factor is given by

$$DF = \frac{1}{(1 + 1i)^{t-0.5}} \qquad (11\text{-}1)$$

where t is the time, and i is the discount rate in fraction.

Profit to Investment Ratio

Profit to investment ratio is the total undiscounted cash flow without capital investment divided by the total investment. Unlike the payout time, it reflects total profitability; however, it does not recognize the time value of money.

Present Worth Net Profit (PWNP)

Present worth net profit is the present value of the entire cash flow discounted at a specified discount rate.

Investment Efficiency or
Present Worth Index or Profitability Index

Investment efficiency or present worth index or profitability index is the total discounted cash flow divided by the total discounted investment.

Discounted Cash Flow Return
on Investment or Internal Rate of Return

Discounted cash flow return on investment or internal rate of return is the maximum discount rate which needs to be charged for the investment capital to produce a break even venture, i.e., the discount rate at which the present worth net profit is equal to zero. This can be also expressed as the discount rate at which the total discounted cash flow excluding investments is equal to the discounted investments over the life of the project.

Scenarios

Economic optimization is the ultimate goal of sound reservoir management. It involves more than one scenario or alternative approach to picking the best solution. For an example, possible choices and questions concerning the recovery scheme and development plan for a waterflood project:

1. Peripheral vs. inside pattern flood
2. Well spacing - number of wells

The economic analyses and comparisons of the results of the various choices can provide the answer to making the best business decision to maximize profits.

Data

The data required for economic analysis can be generally classified as production, injection, investment and operating costs, financial, and economic data. Table 11-1 provides a list of the pertinent data.

Table 11-1 Economic Data

Data	Source/Comment
Oil and gas production rates vs time	Reservoir engineers/Unique to each project
Water injection vs time	Reservoir engineers/Unique to each project
Oil and gas prices	Finance, economic, and strategic planning professionals
Capital investment (tangible intangible) and operating costs	Facilities, operations and engineering professionals/Unique to each project
Royalty/production sharing	Unique to each project
Discount and inflation rates	Finance and economics profession als/Strategic planning interpretation
State and local taxes (production, severance, ad valorem, etc.)	Accountants
Federal income taxes, depletion and amortization schedules	Accountants

Economic Evaluation

The five waterflood design cases presented in chapter 7 were analyzed to determine the potentially most economically viable project.

The procedure used for economic calculation before federal income tax (BFIT) is outlined below:

1. Calculate annual revenues using oil and gas sales from production and unit sales prices
2. Calculate year by year total costs including capital investments (drilling, completion, facilities, and abandonment, etc.), operating expenses and production taxes
3. Calculate annual undiscounted cash flow by subtracting total costs from the total revenues
4. Calculate annual discounted cash flow by multiplying the undiscounted cash flow by the discount factor at a specified discount rate

Table 11-2 presents economic parameters used.[7] The computational procedure is illustrated in a spreadsheet calculation for Case 2 in Table 11-3.

Table 11-2 Economic Parameters

	Case-1	Case-2	Case-3	Case-4	Case-5
Injection Wells	4	8	4	9	13
Production Wells	5	9	1	4	12
Maximum Water Injection					
Total, BWPD	3620	9443	2548	10769	17429
Rate, BWPD/Well	905	1180	637	1197	1340
Maximum Oil Production					
Total, STBOPD	663	1798	606	893	2417
Rate, STBPD/Well	133	200	606	223	201
Maximum Water Production					
Total, BWPD	788	3173	110	1022	6617
Rate, BWPD/Well	158	353	110	255	551
Maximum Gas Production					
Total, MSCFPD	814	4412	497	1981	4259
Rate, MSCFPD/Well	163	490	497	495	355

Depth, ft	5325	Reservoir Pressure, psi	2332
Avg. Thickness, ft	55	Avg. Effective Permeability, md	226
Oil Gravity °API	33	Oil Viscosity, cp	4.8

Reservoir Temperature, °F	123
Gas Gravity	0.67(Air=1)

Table 11-2 Economic Parameters (Contd.)

	Case-1	Case-2	Case-3	Case-4	Case-5
Production Wells					
Old	5	5	1	2	0
New		4		2	12
Injection Wells					
Old	4	4	4	3	5
New		4		6	8
Drilling Costs for Project, $ Million	0.000	0.880	0.000	0.880	2.200
Completion Costs for Project, $ Million	1.345	2.537	0.733	1.917	3.721
Facility Costs for Project, $ Million	0.400	1.261	0.180	0.531	2.578
Abandonment Cost, $ Million	0.108	0.204	0.060	0.156	0.300
Capital Investments	1.853	4.882	0.973	3.484	8.799
Annual Operating Costs, $ Thousand					
Production Wells	110	198	22	88	264
Injection Wells	60	120	60	135	195
Total Annual Project	170	318	82	223	459

Operating Costs, $ Million

Oil Price per STBO, $ 19.5   Gas Price per MCF, $ 2.10

Discount Rate, 12%   Royalty, Severance, and Advolarem taxes 20% of Revenue

Per Well Costs, $ Thousands

Drilling   110   Production Well Completion   53   Injection Well Completion   45

Abandonment   12

Table 11-3 Economic Evaluation - Case 2

	(1)	(2)	(3)	(4)	(5)	(6)	(7)	(8)
Year	Oil Prod. (MSTB)	Oil Price ($/STB)	Oil Revenue ($MM) (1)x(2)/1000	Gas Prod. (MMSCF)	Gas Price ($/MSCF)	Gas Revenue ($MM) (4)x(5)/1000	Total Revenue ($MM) (3)+(6)	Producing Tax ($MM) (7)xTax Rate
1997		19.50	0.00		2.10	0.000	0.000	0.000
1998	275.21	19.50	5.37	1610.5	2.10	3.382	8.749	1.750
1999	189.7	19.50	3.70	503.02	2.10	1.056	4.755	0.951
2000	451.54	19.50	8.81	89	2.10	0.187	8.992	1.798
2001	656.27	19.50	12.80	74.12	2.10	0.156	12.953	2.591
2002	605.71	19.50	11.81	69.86	2.10	0.147	11.958	2.392
2003	515.63	19.50	10.05	60.27	2.10	0.127	10.181	2.036
2004	460.57	19.50	8.98	54.43	2.10	0.114	9.095	1.819
2005	407.65	19.50	7.95	48.8	2.10	0.102	8.052	1.610
2006	351.97	19.50	6.86	42.8	2.10	0.090	6.953	1.391
2007	305.06	19.50	5.95	37.63	2.10	0.079	6.028	1.206
2008	235.47	19.50	4.59	29.77	2.10	0.063	4.654	0.931
2009	227.9	19.50	4.44	28.89	2.10	0.061	4.505	0.901
2010	227.89	19.50	4.44	28.89	2.10	0.061	4.505	0.901
2011	227.9	19.50	4.44	28.89	2.10	0.061	4.505	0.901
Total	5138.47		100.200	2706.87		5.684	105.885	21.177

	(9)	(10)	(11)	(12)	(13)	(14)	(15)	(16)	
Year	Capital Investment ($MM)	Discount Factor @ 12%	Discounted Capital Investment ($MM) (9)*(10)	Operating Cost ($MM)	Total Cost ($MM) (8)+(9)+(12)	Undiscounted Cash Flow ($MM) (7)-(13)	Discounted Cash Flow @ 12% ($MM) (10)*(14)	Cumulative Discounted Cash Flow @ 12% ($MM)	Time (years)
1997	4.678	0.9449	4.420		4.678	-4.678	-4.420	-4.420	1
1998		0.8437	0.000	0.318	2.068	6.681	5.636	1.216	2
1999		0.7533	0.000	0.318	1.269	3.486	2.626	3.842	3
2000		0.6726	0.000	0.318	2.116	6.876	4.624	8.467	4
2001		0.6005	0.000	0.318	2.909	10.044	6.032	14.498	5
2002		0.5362	0.000	0.318	2.710	9.248	4.959	19.457	6
2003		0.4787	0.000	0.318	2.354	7.827	3.747	23.204	7
2004		0.4274	0.000	0.318	2.137	6.958	2.974	26.178	8
2005		0.3816	0.000	0.318	1.928	6.123	2.337	28.515	9
2006		0.3407	0.000	0.318	1.709	5.245	1.787	30.302	10
2007		0.3042	0.000	0.318	1.524	4.504	1.370	31.673	11
2008		0.2716	0.000	0.318	1.249	3.405	0.925	32.598	12
2009		0.2425	0.000	0.318	1.219	3.286	0.797	33.395	13
2010		0.2165	0.000	0.318	1.219	3.286	0.711	34.106	14
2011	0.204	0.1933	0.039	0.318	1.423	3.082	0.596	34.702	15
Total	4.882		4.460	4.452	30.511	75.374	34.702		

Producing Tax Rate (%) = 20.00

	Interest Rate (%)	Discounted Cash Flow ($MM)
Value for Example Above =	12.00	34.702
Starting Interest Rate =	10.00	38.954
	20.00	22.781
	30.00	14.455
	40.00	9.675
	50.00	6.699
	60.00	4.728
	70.00	3.360
	80.00	2.373
	90.00	1.640
	100.00	1.081
	110.00	0.646
	120.00	0.302
	130.00	0.026
	140.00	-0.199
	150.00	-0.384
	160.00	-0.537
	170.00	-0.665
	180.00	-0.773
	190.00	-0.864
Ending Interest Rate =	200.00	-0.942

Payout Time (years) = 1.78
Profit-to-Investment Ratio = 16.44
Present Worth Net Profit ($MM) = 34.702
Present Worth Index = 7.781
Discounted Cash Flow Return on Investment (%) = 131.15

Table 11-4 Economic Evaluation

	Case-1	Case-2	Case-3	Case-4	Case-5
Capital Investment, $MM	1.853	4.882	0.973	3.484	8.799
Reserves, MMSTBO	1.965	5.138	1.378	3.176	5.105
Project Life, Years	15	15	15	15	15
Payout, Years	2.58	1.78	2.44	2.74	2.28
Discounted Cash Flow Return on Investment, %	69.64	131.15	80.12	87.83	104.84
Present Worth Net Profit, $MM	9.454	34.702	7.013	18.721	35.184
Profit-to-Investment Ratio	16.88	16.44	23.32	13.91	8.74
Present Worth Index	5.66	7.78	8.02	5.90	4.35
Development Costs, $/STBO	0.94	0.95	0.71	1.10	1.72

The results of the economic analysis for the 5 waterflood cases are presented in Table 11-4, which shows that all the cases are very favorable for waterflooding. Note that federal income taxes are not taken into account.

Case 3 gives the lowest amount of investment, reserves, and development costs and yet very favorable discounted cash flow return on investment, and the highest profit-to-investment-ratio. Case 2 for peripheral flood and Case 5 for pattern flood show the most promise. Case 5 shows the highest present worth net profit; however, this requires 80% more capital than for the Case 2, which gives about the same present worth net profit and better discounted cash flow return on investment.

It should be realized that for the Case 2 peripheral flood, recovery estimates may be optimistic because the reservoir layers were considered to be homogeneous and continuous. This may not represent the real situation. On the other hand, Case 5 for the pattern flood case is better suited to treat reservoir heterogeneity and reservoir discontinuity.

The selection of the optimum case will depend on availability of capital, technical consideration, and the risk involved. The sensitivity analysis discussed below shows that Case 2, even if recovery is 20% lower, can still be the best choice.

Risk and Uncertainties

The very nature of economic evaluation entails risk taking and uncertainties involving technical, economic, and political conditions. The results of the analysis are subjected to many restrictive assumptions in forecasting recoveries, oil and gas prices, investment and operating costs, and inflation rate. Unforeseen national and world economic and political climates can also severely affect the outcome of the projects.

Figure 11-2 shows the sensitivities of DCFROI and PWNP to the oil price, oil production, investment, and operating costs. The analysis shows that DCFROI is affected more drastically by oil price, oil production, and investment than by the operating costs. PWNP is most sensitive to oil price and oil production.

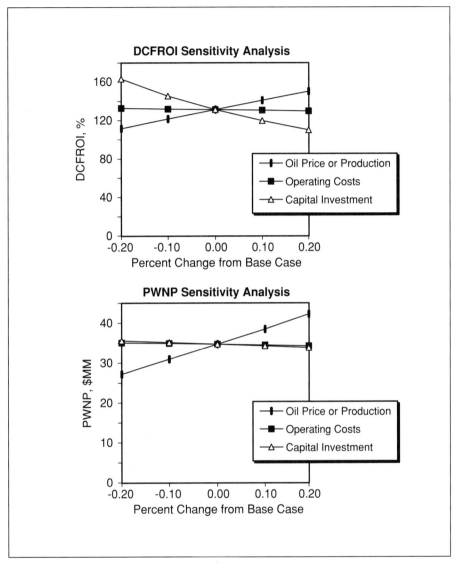

Figure 11-2 Sensitivity Analysis of Case 2

References

1. Satter, A., and G. C. Thakur. "Integrated Reservoir Management: A Team Approach," PennWell Books, Tulsa, OK(1994).
2. Stermole, F. J. and J. M. Stermole. "Economic Evaluation and Investment Decision Methods," Sixth Ed., Investment Evaluation Corporation, Golden, CO, 1987.
3. Seba, R. D. "Determining Project Profitability," *JPT* (March 1987) 263-71.
4. Garb, F. A. "Assessing Risk In Estimating Hydrocarbon Reserves and in Evaluating Hydrocarbon-Producing Properties," *JPT* (June 1988) 765-68.
5. Rose, S. C., J. F. Buckwalter, and R. J. Woodhall. "The Design Engineering Aspects of Waterflooding," SPE Monograph 11, Richardson, TX, 1989.
6. Hickman, T. S. "The Evaluation of Economic Forecasts and Risk Adjustments in Property Evaluation in the U.S.," *JPT* (February 1991) 220-25.
7. Jones, C. Personal communication, Texaco, Inc.

12

Case Studies

The purpose of this chapter is to document best practices involving integrated waterflood asset management for onshore and offshore fields. Several case studies outlining the benefits of working as cross-functional teams are discussed in this chapter:

- Elk Basin Madison
- Denver Unit
- Means San Andres Unit
- Jay/LEC Fields
- Ninian Field

Elk Basin Madison[1]

The Elk Basin Madison reservoir provides an example of reservoir's description as an iterative process, and illustrates the importance of obtaining extensive reservoir data during field development so that reservoir geology can be defined as soon as possible and incorporated into waterflooding plans.

During the Elk Madison waterflood, a revised reservoir description was used to help interpret the observed production data. In the initial water injection program, water breakthrough was rapid in the interior wells and caused scaling problems, which resulted in production rate declines. In combination with these initial results, a revised description was utilized to alter the water injection program and to drill new producing wells in underdeveloped areas. Figure 7-1 shows the performance history of the Elk Basin Madison resulting from this analysis. This results in an increase in ultimate recoverable reserves of 62 million barrels, or 8% OOIP.

The Madison reservoir in the Elk Basin Anticline is shown in Figures 12-1 and 12-2. Performance data in the field shows a small pres-

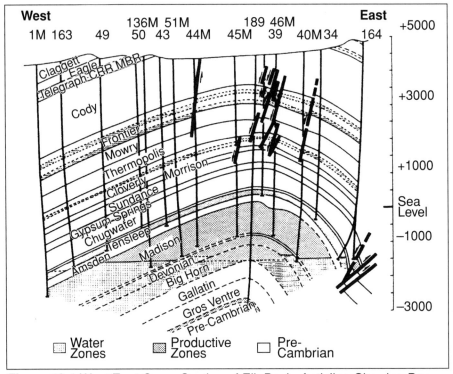

Figure 12-1 West/East Cross Section of Elk Basin Anticline Showing Pay Horizons (Copyright 1969, SPE, from JPT, February 1969[1])

sure decline, indicating a strong water drive. The reservoir was considered homogeneous with a high level of connectivity between the lower and upper areas. As a result, the entire section was considered to be under active water drive.

Based upon the above concept, a homogeneous reservoir model was developed. However, as development proceeded, interpretation of well logs led to a definition of the reservoir with four distinct zones A, B, C, and D. Even at this stage the four zones were assumed to be under active water drive, although non-communicating. Later on when new wells were drilled in separate zones, it was realized that Zone A did not have any water influx and had a low reservoir pressure; whereas Zones B, C and D responded as expected under a water drive. Increased fluid withdrawals caused an increase in GOR in Zone A and a decrease in production rates.

As a result of extensive reservoir characterization, utilizing core, log, and production data. About 3,000 ft of core was available in 10

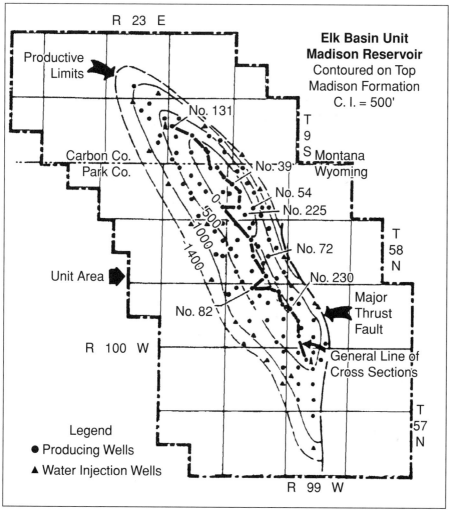

Figure 12-2 Plat Showing Elk Basin Madison Structure and Location of Wells (Copyright 1969, SPE, from JPT, February 1969[1])

wells. Figures 12-3 and 12-4 show reservoir zonation interpreted from core and log data. The zones show a high degree of variability in permeability and vertical/lateral continuity. Zone A possesses a high permeability, with low lateral continuity of the pay zones and a lack of a natural water influx. On the other hand, Zones B, C, and D are characterized by lower permeability, a higher degree of lateral continuity, and a stronger natural water influx.

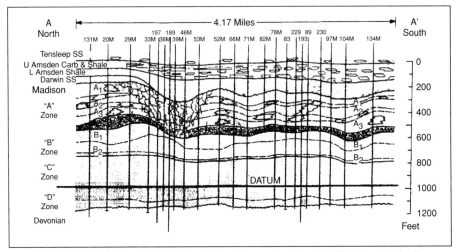

Figure 12-3 North/South Stratigraphic Cross Section of Elk Basin Madison Reservoir Showing Karst Development, a Collapsed Sinkhole on the North End, and Lateral Continuity (Copyright 1969, SPE Textbook Series Vol. 3)

Table 12-1 provides a summary of the zone characteristics. This revised description was helpful in interpreting the observed production data. The pressure in Zone A declined rapidly, whereas other zones maintained pressures as a result of water drive.

Table 12-1 Summary of Elk Basin Madison Zone Characteristics (Copyright 1969, SPE, from JPT, February 1969[1])

Zone	Average Net Pay (ft)	Gross Pay Sections (ft)	Porosity (%)	Permeability (md)	Lithology of Pay	Recovery Mechanism
A₁ₐ	8	25	13.6	48		
A₁ᵦ	10	20	12.1	71	Fine- to medium-grained dolomite	
A₂ₐ	19	38	12.4	47	Medium-grained dolomite	Solution gas drive artificial waterflood
A₂ᵦ	9	45	11.3	73	Medium-grained dolomite with vugs	
A₃	22	62	12.1	368		
					Solution Breccia Zone	
B₁ₐ	37	65	11.5	5		
B₁ᵦ	22	100	10.0	3	Fine crystalline dolomite	
B₂	6	70	11.5	6		Natural water drive
C	7	215	10.3	3	Microcrystalline dolomite	
D	48	140	11.2	11	Fine crystalline dolomite	

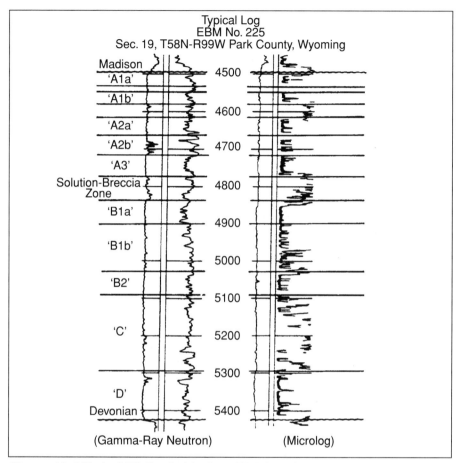

Figure 12-4 Typical Elk Basin Madison Well Log Showing Subzones (Copyright 1969, SPE, from JPT, February 1969[1])

A peripheral water injection was initiated in Zone A, and water injection in lower zones was discontinued. Also, new production wells were drilled in the underdeveloped areas. As a result, an increase in production of about 10,000 BOPD was observed, leading to an increase in ultimate reserves of 62 MMBO, or 8% OOIP.

This example illustrates the importance of characterizing the reservoir early on so that waterflooding can be implemented effectively as soon as practical.

Denver Unit Waterflood[2-6]

Field Discovery and Development

The Denver Unit waterflood in the Wasson San Andres field of west Texas, Figure 12-5, shows how geological concepts were used to redesign a waterflood. The field was discovered in 1936, and it produces from the San Andres carbonated formation at a depth of about 5,000 ft. The formation thickness varies from 300 to 500 ft. The primary producing mechanism was solution gas drive. Primary development was on 40-acre spacing.

Secondary operation. Secondary recovery utilizing peripheral water injection started in 1964 (see Figure 12-6). Water was injected below the oil-water contact, as shown in Figure 3-14. The reservoir was considered continuous vertically and laterally, and thus believed that the injected water from below the oil-water contact would displace oil effec-

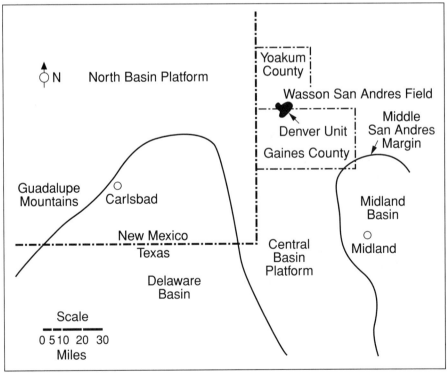

Figure 12-5 Location Map-Denver Unit, Wasson San Andres Field (Copyright 1974, SPE, from JPT, June 1974[5])

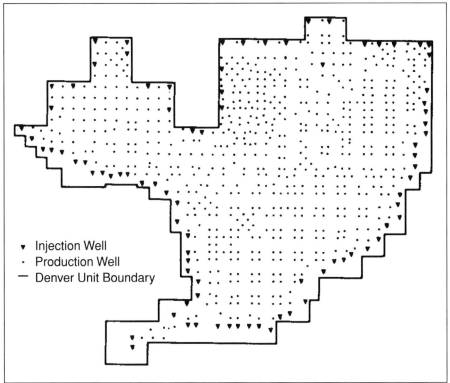

Figure 12-6 Original Peripheral Waterflood Pattern (Copyright 1980, SPE, from JPT, September 1980[4])

tively. However, the performance of the waterflood was less than expected. Water injectivity was low because of peripheral injection wells injecting less water as a result of poor reservoir rock quality.

Typically, the first row of production wells away from the injection wells showed some response; however the production wells located several miles away did not see any response at all. Poor reservoir continuity and transmissibility coupled with low water injectivity caused the peripheral waterflood to fail (see Figure 12-7). As a result, a pattern waterflood was started (see Figure 12-8).

The reservoir characterization concept of the Denver Unit changed during and as a result of the peripheral waterflood. A detailed geologic study indicates that the total vertical section was made up of 10 distinct zones (see Figure 12-7). Various zones were mapped laterally and areally, with barriers identified between zones. Also, it was conclud-

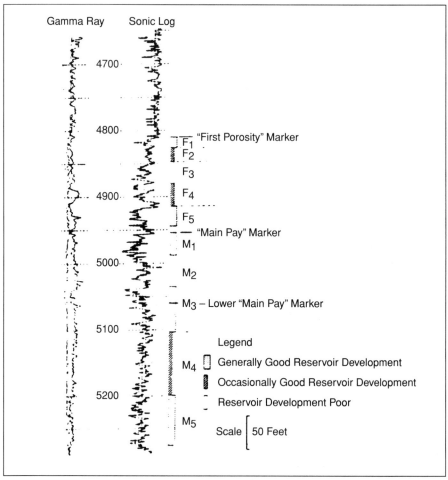

Figure 12-7 Zonal Subdivisions of San Andres Reservoir (Copyright 1974, SPE, from JPT, June 1974[5])

ed that some zones were not continuous over a long distance. This explained the ineffective waterflood performance on a 40-acre spacing.

Further geologic and engineering led to infill drilling on 20-acre spacing, and an inverted 9-spot pattern, with one injection well for every three production well, was developed (Figure 12-8). Production performance from the waterflood is shown in Figure 12-9.

The performance curves, as shown in Figure 12-9, show a decreasing GOR, increasing water injection rate, reservoir voidage and oil production rate. These performance characteristics clearly indicate the successful performance of the waterflood.

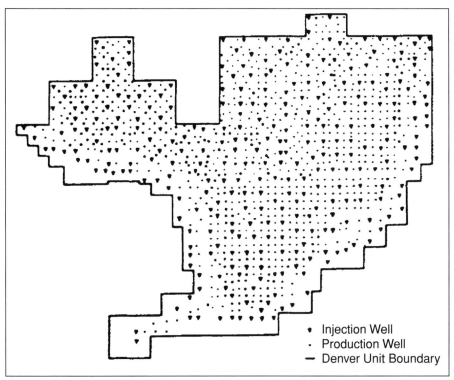

Figure 12-8 1979 Project Status (Copyright 1980, SPE, from JPT, September 1980[4])

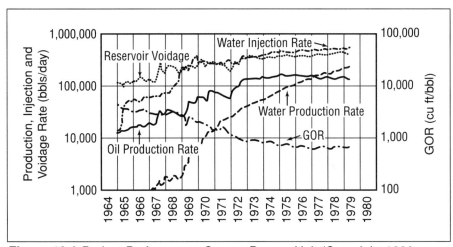

Figure 12-9 Project Performance Curves-Denver Unit (Copyright 1980, SPE, from JPT, September 1980[4])

Means San Andres Unit[7]

Introduction

This case study provides information on the evolution of reservoir management to meet changing economic and technical challenges as the field produced by primary, secondary, and tertiary methods. Reservoir management at Means San Andres Unit has consisted of an ongoing but changing surveillance program supplemented with periodic major reservoir studies to evaluate and make changes to the depletion plan.

Reservoir management techniques began within one year of discovery and became increasingly complex as operations changed from primary to secondary to tertiary. Reservoir description, infill drilling with pattern modification, and reservoir surveillance played key roles in reservoir management. Reservoir description methods evolved from the relatively simple techniques used in the 1930s to the recent use of high-resolution seismic to improve pay correlation between wells.

Field Discovery and Development

The Means field was discovered in 1934. It is located about 50 miles northwest of Midland, Texas, along the eastern edge of the Central Basin platform (see Figure 12-10). Table 12-2 lists the reservoir and fluid properties. The field is a north-south-trending anticline separated into a North Dome and a South Dome by a dense structural saddle running east and west near the center of the field (see Figure 12-11). Production in this field is from the Grayburg and San Andres formations at depths ranging from 4,200 to 4,800 ft. Figure 12-12 is a type log, showing the zonation.

The Grayburg is about 400 ft thick with the basal 100 to 200 ft. considered gross pay. Production from the Grayburg was by solution-gas drive with the bubble-point at the original reservoir pressure of 1,850 psi. The Grayburg is of poorer quality, and its production has been minor relative to the San Andres. The San Andres is more than 1,400 ft. thick, with the upper 200 to 300 ft. being productive. The primary producing mechanism in the San Andres was a combination of fluid expansion and a weak waterdrive.

Table 12-2 Reservoir and Fluid Properties, Means San Andres Unit
(Copyright 1992, SPE, from JPT, April 1992)

Formation name	San Andres
Lithology	Dolomite
Area, acres	14,328
Depth, ft	4,400
Gross thickness, ft	300
Average net pay, ft	54
Average porosity, %	9.0 (up to 25)
Average permeability, md	20.0 (up to 1,000)
Average connate water, %	29
Primary drive	Weak water drive
Average pressure (original), psig	1,850
Stock-tank gravity, °API	29
Oil viscosity, cp	6
FVF, RB/STB	1.04
Saturation pressure, psi	310

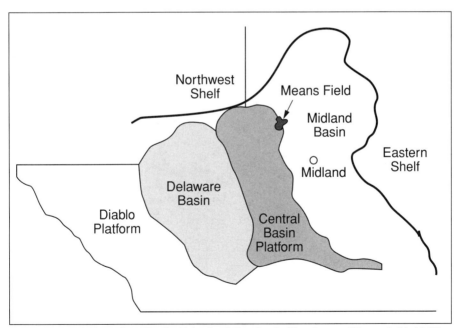

Figure 12-10 Permian Basin Principal Geologic Provinces (Copyright 1992, SPE, from JPT, April 1992[7])

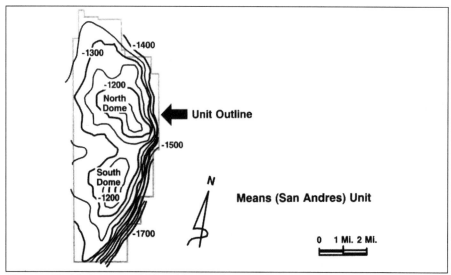

Figure 12-11 Structural Map-Means San Andres Unit (Copyright 1992, SPE, from JPT, April 1992[7])

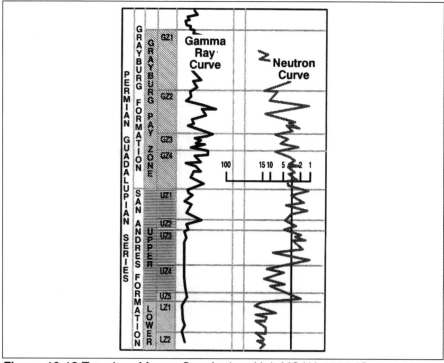

Figure 12-12 Type Log Means-San Andres Unit MSAU 1216 (Copyright 1992, SPE, from JPT, April 1992[7])

Reservoir Management During Primary and Secondary Operations

The first reservoir study was completed in 1935, concentrating on reservoir management related to primary recovery. Later, in 1959, a reservoir study was conducted to evaluate secondary recovery. Highlights of this study included one of Humble's first full-field computer simulations. For this study, additional data were accumulated, including further loggings, fluid sampling, and special core data (e.g., capillary pressures and relative permeabilities).

Cross-sections, like the one shown in Figure 12-13, aided in the design of an initial waterflood pattern. In 1963, the field was unitized and a peripheral flood was initiated (see Figure 12-14). Twenty-four wells, distributed throughout the unit, were permanently shut-in and maintained as pressure response wells to monitor reservoir pressure. In 1967, as a result of increased allowable, it was realized that the peripheral injection pattern no longer provided sufficient pressure support.

In 1969, a reservoir engineering and geological study was conducted to determine a new depletion plan to offset the pressure decline. The geologic study included a facies study. In the North Dome, pressure data were correlated with the geological data to identify three major San Andres intervals: Upper San Andres, Lower San Andres oil zone, and Lower San Andres aquifer. A permeability barrier was mapped between the Upper and Lower San Andres. Analysis of pressure data from the observation wells indicated that neither North Dome nor South Dome were receiving adequate pressure support. This study recommended interior injection with a 3:1 line drive (see Figure 12-15). Following implementation of this program, the unit production increased from 12,000 BOPD in 1970 to greater than 18,000 BOPD in 1972.

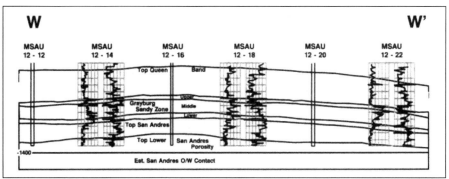

Figure 12-13 West-East Structural Cross Section-Means San Andres Unit (Copyright 1992, SPE, from JPT, April 1992[7])

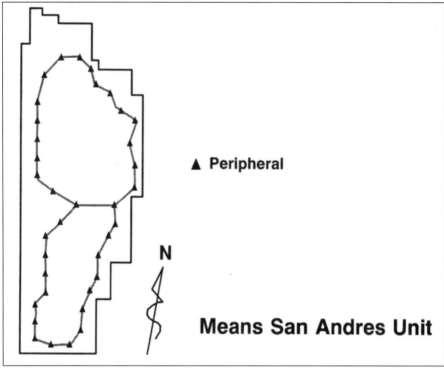

Figure 12-14 Waterflood Injection Pattern (Copyright 1992, SPE, from JPT, April 1992[7])

After reaching a peak in 1972, oil production again began to decline. A reservoir study conducted in 1975 indicated that all the pay was not being flooded effectively by the 3:1 line-drive pattern. An in-depth geological study showed a lack of lateral and vertical distributions of pay. Old gamma-ray/neutron logs were correlated with core data to determine porosity-feet. Original oil-in-place was calculated for up to six zones in each well in the field. This geological work provided the basis for a secondary surveillance program and later for the design and implementation of the CO_2 project.

A major infill drilling program was undertaken based upon potential additional recovery estimated from infill drilling with pattern densification (see Figure 12-16). From 141 (20-acre spacing with an 80-acre inverted 9-spot pattern) infill wells, about 15.4 million BO of incremental oil was estimated (see Figure 12-17 for oil production with and without infill program). With the implementation of this pattern flooding, a detailed surveillance program was developed, including:

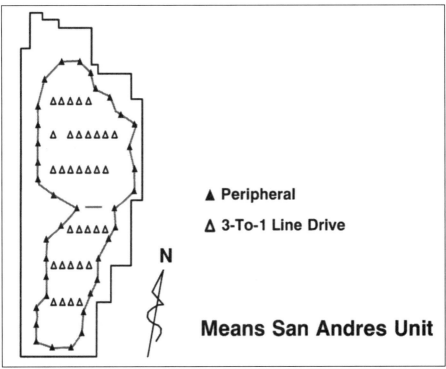

Figure 12-15 Waterflood Injection Pattern (Copyright 1992, SPE, from JPT, April 1992[7])

- monitoring of production (oil, gas and water)
- monitoring of water injection
- control of injection pressures with step-rate tests
- pattern balancing with computer balance programs
- injection profiles to ensure injection into all pay
- specific production profiles
- fluid level checks to ensure pump-off of producing wells

Reservoir Management During Tertiary Operations

In 1981-1982, a CO_2 tertiary recovery reservoir study was conducted. At that time, several major CO_2 projects had been proposed for San Andres reservoirs, but none had been implemented. Although Means was similar to other San Andres fields in the Permian Basin,

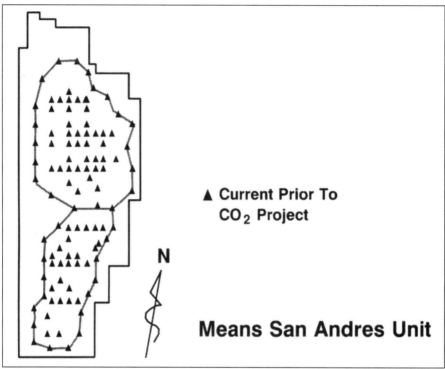

Figure 12-16 Waterflood Injection Pattern (Copyright 1992, SPE, from JPT, April 1992[7])

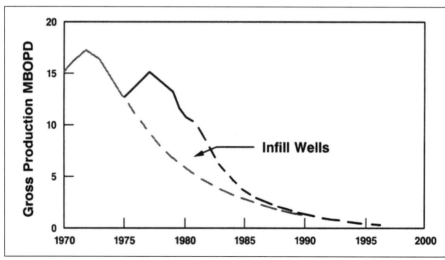

Figure 12-17 MSAU Infill Drilling (Copyright 1992, SPE, from JPT, April 1992[7])

some properties, e.g., 6-cp oil viscosity, relatively high minimum miscibility pressure, and low formation parting pressure, made the Means Unit somewhat unique.

A CO_2 flood pilot along with extensive laboratory and simulation works were initiated. A detailed reservoir description program preceded this work and became the basis for planning the CO_2 tertiary project. Although this reservoir description was the building block for the project, it was continuously updated during the planning and implementation phases of the CO_2 project as more data became available.

Several 10-acre wells (generally injectors) were drilled as a part of the CO_2 project. The project consisted of 167 patterns on 6,700 acres, and it included 67% of the productive acres and 82% of OOIP.

A comprehensive surveillance program had been present during the waterflood. Before developing a similar program for CO_2 flooding, an operating philosophy was created by personnel from engineering, geology, and operations and submitted to management for approval and support. Major operation objectives included:

- completing injectors and producers in all floodable pay
- maintaining reservoir pressure near the MMP of 2,000 psi
- maximizing injection below fracture pressure
- pumping off producers
- obtaining good vertical distribution of injection fluids
- maintaining balanced injection/withdrawals by pattern

To satisfy these objectives, a surveillance program was developed involving engineers, geologists and operations personnel. Major areas of surveillance included:

- areal flood balancing
- vertical conformance monitoring
- production monitoring
- injection monitoring
- data acquisition and management
- pattern performance monitoring
- optimization

The objectives of the above surveillance program were to maximize oil recovery and flood efficiency, and to identify and evaluate new opportunities and technologies. In addition, one of the key objectives was to obtain better reservoir descriptions to understand the reservoir processes. This effort included the use of high resolution seismic to improve pay correlation between wells. The application of seismic sequence stratigraphic yielded significant insights into reservoir's complexities. It provided geometric template to constrain basic well-log correlations properly within a chronostratigraphic framework. Seismic-scale stratal geometries, combined with detailed geologic rock descriptions, defined this sequence framework. In addition, the stratigraphic model was used to optimize CO_2 project completions by ensuring that all pays are opened in both producers and injectors based on correlative stratigraphic zones.

As mentioned earlier, the CO_2 project was implemented as part of an integrated reservoir management plan, including CO_2 injection, infill drilling, pattern changes, and expansion of the Grayburg waterflood outside the project area. Total unit oil production is shown in Figure 12-18 with and without the project. Figure 12-19 shows the performance of the tertiary project area.

Jay/LEC Fields[8-11]

Jay-Little Escambia Creek waterflood provides one of the best documented examples of acting on available information. It illustrates how an extensive geological study performed during development of the field was integrated into the waterflood design for a major project.

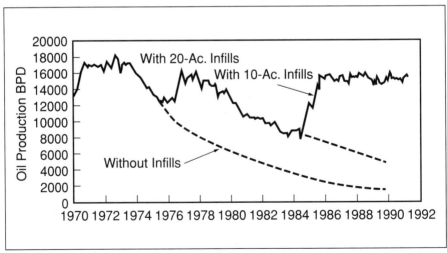

Figure 12-18 Means San Andres Unit Oil Production 1970-1990 (Copyright 1992, SPE, from JPT, April 1992[7])

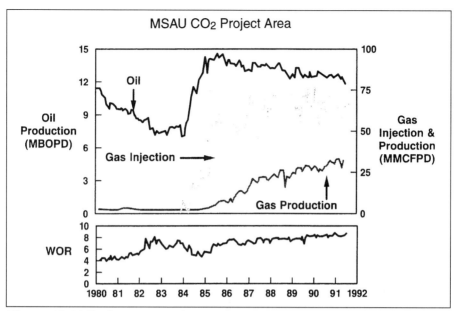

Figure 12-19 Performance of Tertiary Project Area (Copyright 1992, SPE, from JPT, April 1992[7])

Field Discovery and Development

The Jay/LEC field was discovered in 1970. The structure map, on the top of Smackover Jay/LEC fields, is shown in Figure 12-20. Production is from the Smackover carbonate and Norphlet sand formations at a depth of about 15,400 ft. Figure 12-21 is a typical well log from the field. More than 90% of OOIP is in the first Smackover formation.

The reservoir is highly undersaturated, with the initial and saturation pressures of 7,850 and 2,830, respectively. Early reservoir performance indicated that a natural water drive would not be an effective source of reservoir energy. Thus, a pressure maintenance program, utilizing water injection, was selected.

Reservoir Characterization and Management During Secondary Operations—Extensive reservoir description data was obtained during field development. One hundred and two development wells were cored between 1970 and 1974. This produced a wealth of reservoir rock data to use in describing the reservoir, and provided a basis for unitization and for planning and justifying a waterflood.

High and low permeability zones were correlatable over a large area of the field, implying that thin, tight streaks with limited or no ver-

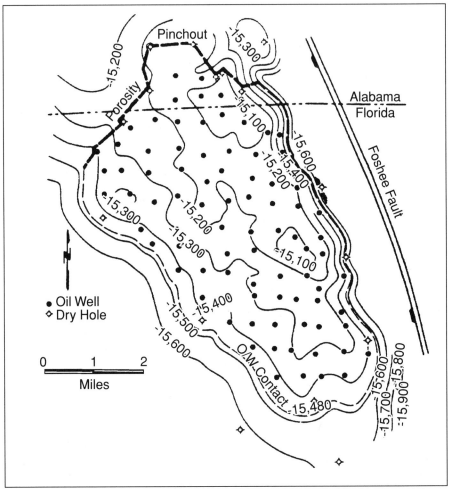

Figure 12-20 Structure Map, Top of Smackover Jay/LEC Fields (Copyright 1983, SPE, from paper 61998)

tical flow capability could inhibit crossflow between layers. The first Smackover formation was selected for waterflood first because of favorable geological correlations. The mobility ratio between water and oil was quite favorable at 0.30.

The waterflooding plan was developed by using a simulation study and taking advantage of available geologic information. Four waterflood plans were considered: peripheral flood, five-spot pattern (Figure 12-22), 3:1 staggered live drive (Figure 12-22), and a combination pattern (Figure 12-23). The combination pattern was chosen as it matched

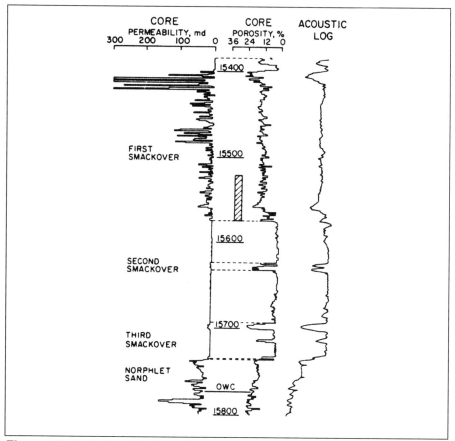

Figure 12-21 Typical Well Log, Exxon-Bray 10-4, Jay/LEC Fields (Copyright 1983, SPE, from paper 61998)

the geology. As seen from Figure 12-23, the thick, central portion of the reservoir was flooded as a 3:1 staggered line drive. Figure 12-24 shows the waterflood performance.

In 1973, as planning for waterflood progressed, the operator realized that attaining the goal of optimum recovery would require augmenting the reservoir description with data from an extensive surveillance program. The resulting program was developed to monitor and control waterflood conformance in the thick-layered reservoir. Cased hole logging, pressure buildup and production tests, and permeability data taken from conventional core analysis, provided the vertical conformance data. The areal conformance surveillance employed radioactive tracers, reservoir pressure data and interference tests.

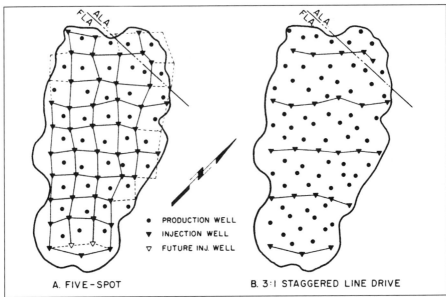

Figure 12-22 Repeating Patterns 2D Areal Reservoir Model, Jay/LEC Fields (Copyright 1983, SPE, from paper 61998)

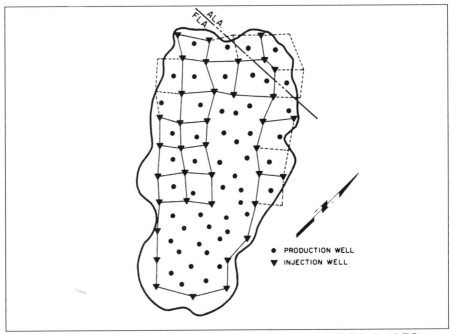

Figure 12-23 Combination Pattern, 2D Areal Reservoir Model, Jay/LEC Fields (Copyright 1983, SPE, from paper 61998)

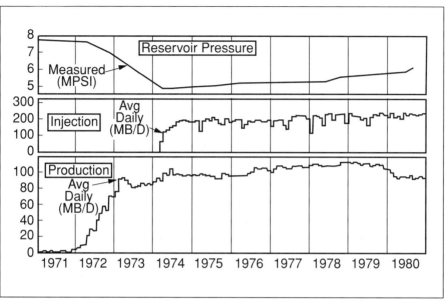

Figure 12-24 Jay/LEC Unit Performance (Copyright 1983, SPE, from paper 61998)

The information produced by combining the surveillance data with the reservoir description data yielded many new insights into water movement and zone depletion. The operator developed new fluid movement analysis techniques and defined several methods for improving recovery, including drilling, workovers, facility modifications and injection balancing. Through these efforts, the operator managed to delay the field's production decline until late 1979. By that time, 75% of the estimated ultimate recoverable reserves had been produced.

In 1977, the waterflood's performance was studied comprehensively, using the data and procedures previously described. One major conclusion was that areal sweep efficiency was generally good due to a very favorable water/oil mobility ratio (0.3:1), but that vertical sweep efficiency was poor in areas of high permeability contrast. The lower-permeability zones not only flooded more slowly, but in some cases showed lower reservoir pressure. Modeling work indicated that the controlling factor was the thick, layered formation.

The study group mapped the areas of upswept oil in each zone, and developed work programs for improving sweep efficiency, which included infill drilling, workovers, injection balancing and surface facility modifications. seventeen infill wells were drilled between May, 1977 and October, 1979. Fourteen of these wells were producers and three were injectors. All were drilled in zones and locations chosen to improve areal and/or vertical conformance. The success of this strategy was outstanding. Additional oil recovery totaled more than 33 million STB.

This project's success prompted further drilling activity. A detailed study completed near the end of 1979 identified 21 additional infill wells. Their locations were chosen in order to recover additional oil from discontinuous zones, improve vertical sweep efficiency by selectively completing in low-permeability zones which were flooding slowly, and boost a real sweep efficiency by locating wells outside the existing waterflood patterns. Also, some of the planned workovers were replaced by new infill wells, since these new wells could be optimally placed and so able to drain additional oil that the workovers could not recover. Further analysis showed that some proposed locations were not economically feasible, but the yield from those that were justified was expected to total 8 million STB of additional oil.

By adding core data from the new wells to their waterfront location maps and their permeability trend interpretations, the operator developed improved well completion techniques. These techniques played an important role in the success of these projects. Two of the most successful techniques were selective completions for zone isolation and preferential stimulation. The selective completion procedure used wellbore casing and cement to maintain zone isolation while opening selected (usually low-permeability) members. This technique was employed successfully in four new well completions. The preferential stimulation technique was designed to minimize reservoir stratification effects and increase recovery from the less-permeable members. In the completion procedure, the low-permeability intervals were opened and stimulated before the higher-permeability sections were performed. The high-permeability intervals were subjected to little or no stimulation after perforation.

Several workovers were performed to improve conformance. The overall success ratio was 50%, and follow-up analysis indicated an increased oil recovery of 5 million STB and a total initial production increase of 8,500 BOPD. This increased recovery resulted from draining low-permeability or plugged undepleted zones in twelve producers and from improving volumetric conformance in three injectors.

The injection balancing effort was designed to optimize volumetric sweep efficiency by effecting waterfront closure in all zones from opposite directions on all producing wells and maintaining balanced zonal pressures to prevent well-bore crossflow. Although this objective was seldom realized in practice, significant results were attained through daily surveillance of injection well rates and weekly monitoring of producing well water cuts. Timely injection rate enabled the operator to maintain flowing production from 18 wells for several months beyond the limit indicated by initial watercut trends. Total increased oil recovery from this procedure was estimated to be 7 million STB.

As the number of active wells in the field declined, many opportunities for flow-line looping and piping modifications developed. The justification for this work was normally based on increased flow rates and additional oil recovery due to lower wellhead pressures and greater wellbore drawdowns. In many cases, surveillance data gave early indications of a low-permeability zone. Here, increased wellbore drawdown improved drainage. Thirty-two modifications were made, resulting in an estimated increased oil recovery of 1 million STB.

All production wells were produced by natural flow. Reservoir pressure was increased to a level near the original pressure of 7,850 psi so that wells could be produced to depletion at essentially 100% watercut. However, flowing wells died at 65 to 75% water cut when produced against the normal treating plan inlet pressure of 275 psi. In several locations, waterfront movement threatened to push oil past the producers, which were nearly watered out. If flow from the producing well was not maintained, substantial volumes of oil would have been lost, migrating to areas of the reservoir where no drainage point existed. Since the presence of this oil could be confirmed from existing waterfront maps, low-pressure (50 psi) systems were installed at two of the treatment plants to sustain flow from the wells in question. The effectiveness of the system was demonstrated by St. Regis 43-2, which had died at 67% water cut prior to system installation. This well was restored to flowing production into the low-pressure system, with a significantly lower wellhead pressure. It began to flow at a rate of 1,000 BOPD, with a 73% water cut.

Ready access to documented surveillance and reservoir description data expedited operations and management decisions. The data provided a current base for decisive action in both day-to-day operations and longer-range planning.

Early diagnosis of well problems was vital in managing this layered reservoir. Establishing permeability profiles for each well and zonal flood front definitions greatly enhanced diagnosis. This was especially true for producing wells with initial or increasing water breakthrough. By defining both the zone(s) and direction of water breakthrough, the operator could begin remedial action promptly.

Forecasting future producing rates and ultimate recovery were also important in effective management. The surveillance tools previously described were used in conjunction with the wealth of reservoir description data to provide an excellent base for short- and long-range forecasting of both individual well and field-wide performance. In addition to the usual crude oil supply and financial planning uses, these forecasts were used in well studies and in planning and proposed tertiary recovery project.

The cost of obtaining core data from every well and conducting a comprehensive surveillance program was substantial. However, a cost/benefits analysis shows that this data acquisition was both practical and profitable.

Field-wide coring during development (from 1970 to 1974) cost about $5 million. Coring the infill wells increased the total cost to $8 million. The total cost for surveillance activities amounted to $3 million. In view of the sizable monetary requirements for reservoir description and surveillance, and in the light of the potential benefits, effective application of the data acquired in these procedures was imperative. Total expenditures for projects generated by these techniques were $90 million.

We have already seen the specific recovery benefits attributable to conventional coring. These benefits were updated and included in the expected total increase in oil recovery–60 million STB–from reservoir management programs. The largest single contribution was from infill drilling which, alone, was expected to increase recovery by 41 million STB. Workovers to improve conformance were estimated to increase recovery by another 11 million STB (including 6 million STB from non-rig work previously attributed to coring). Injection balancing and facility modification were expected to add 7 million STB and 1 million STB, respectively. From these cost and benefits data, we can see clearly that the increased oil recovered through this program cost less than $2 per barrel.

Effective reservoir management also maintained field production capacity above 100,000 barrels per day and (as we have seen) delayed production decline until late 1979, after 75% of the estimated ultimate recovery had been produced. Cumulative oil production from the field amounted to 296 million barrels as of January, 1981 (which was 86% of total reserves). Production averaged 90,000 barrels per day from 31 wells, and it appeared that the remaining reserves of 50 million STB could be recovered readily by continuing waterflood operations. Current recovery plans include supplementing water injection with nitrogen gas, which is miscible with the reservoir crude. This technique is expected to recover an additional 50 million STB over and above projected waterflood recovery.

Some of these benefits could have been achieved without the comprehensive surveillance and reservoir description data previously illustrated. But it is doubtful that much of this work could have been justified or economically supported without these data.

Ninian Field[12]

Introduction

This case study provides information on an offshore waterflood, in which downdip peripheral water injection was adapted in the beginning and later it was augmented by a modified 5-spot pattern to allow higher processing rates. Improving sweep efficiency and effective well management played key roles in improving oil recovery.

Water injection started six months after production startup because of absence of adequate aquifer support for reservoir withdrawals. The field was operated with a balanced injection-withdrawal to maintain the reservoir pressure around 5,000 psi so that wells would continue to flow at high water cuts without artificial lift.

Field Discovery and Development—The Ninian field is located in Blocks 3/3 and 3/8a in the UK North Sea, 100 miles off the Shetland Isles, Scotland. The production started in 1978. Peak oil production was achieved in 1982 with a rate of 307 MBOPD. The field contained OOIP of about 2.6 billion BO in the Brent Sandstone group of Middle Jurassic age.

Some of the reservoir's parameters are given in Table 12-3. Figure 12-25 is a structure map of the top Brent surface, and Figure 12-26 describes the reservoir zonation. There is a shale break with Zone E (Figure 12-26). The upper reservoir is above the break, and the lower reservoir is below the break.

Reservoir Characterization

Forty four wells have been cored and analyzed. Detailed reservoir characterization has been performed, and the field is more heterogeneous than originally envisaged. Consequently, water breakthrough occurred much earlier than anticipated and the buildup in water cut has also occurred more quickly than projected.

Table 12-3 Reservoir and Fluid Properties, Ninian Field (Copyright 1996, SPE, from JPT, December 1996[12])

Formation name	Brent
Lithology	Sandstone
Depth, ft	9,750
Temperature, °F	215
Porosity, %	16-23
Primary drive	Weak water drive
Initial reservoir pressure, psi	6,500
Oil gravity, °API	36
Bubble point pressure, psi	1,250
OOIP, MMMBO	2.6
Production started	1978
Peak oil production rate, MBOPD	307 (in 1982)
Water injection started	6 months after production startup
Cumulative oil production, MMMBO	1.0 (in 1995)
Cumulative water injection, MMMBW	3.0 (in 1995)
Injection rate, MBWPD	600 (in 1995)
Production rate, MBWPD	550 (in 1995)
Oil production rate, MBOPD	60 (in 1995)

Sweep Efficiency Improvements

The following practices have been followed to improve sweep efficiency:

- Use dip to improve sweep efficiency for water injection
- Utilize selective completion, with four packers and pipeline sliding sleeve/plug operations, to produce from zones individually or in a number of combinations (Figure 12-27)
- Apply remedial well work, including perforating new intervals, reconfiguring the completions, squeeze cementing perforations
- Perforate smaller intervals (<10 ft)
- Use ESPs (electrical submersible pumps) to increase drawdown in some wells
- Improve areal sweep by continuous realignment of the waterflood, including converting to more updip injection wells

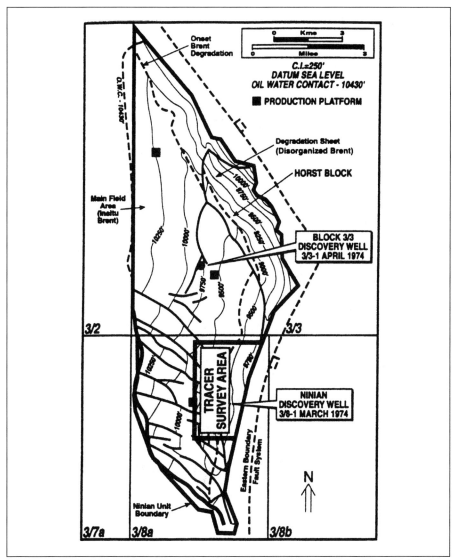

Figure 12-25 Ninian Field Top Brent Structure Map (Copyright 1996, SPE, from JPT, December 1996[12])

Well Management To Improve Recovery

Good well-management practices have resulted in effective reservoir management in the Ninian field:

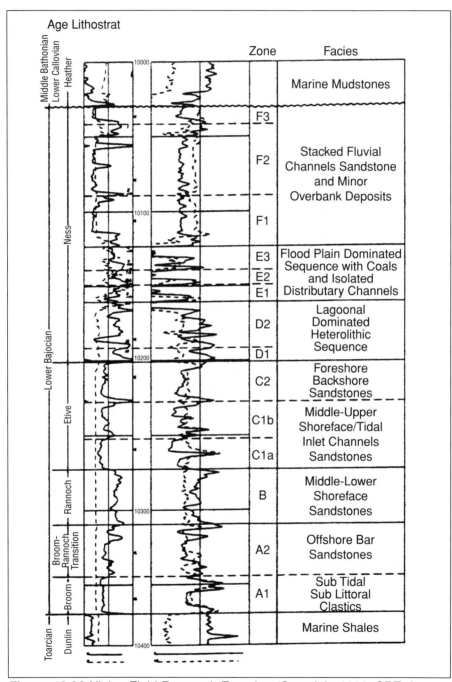

Figure 12-26 Ninian Field Reservoir Zonation (Copyright 1996, SPE, from JPT, December 1996[12])

Perf MD	Well Schematic	MD	Description	Maker/Type	Min. I.D.
		679	Tree Number E228 TRFV	Otis Series 10	4.562"
			Tubing (300 JTS) 5–1/2" 17 PPF	FOX L80 13 CR	4.892
		3478	Landing Nipple	Otis XN	4.157
		12365	Crossover 5–1/2" x 4–1/2"		3.958
		14332	Landing Nipple	Otis X	3.813
		14385	P.B.R.	Baker	3.974
		14385	Anchor	Baker K-22	3.875
		14411	Permanent Packer	Baker FAB	4.000
		14414	Millout Extension		4.408
		14818	P.B.R.	Baker	3.875
		14846	Anchor	Baker K-22	3.875
		14550	Permanent Packer	Baker FAB	4.000
		14555	Millout Extension		4.408
			Tubing (52 JTS) 4–1/2" 12.6 PPF	FOX L80 13 CR	3.956
14500-14628		14812	Sliding Sleeve (Open)	Otis RD	3.486
		14616	Tubing Fill		
		14634	Anchor	Baker K-22	3.875
		14635	Permanent Packer	Baker FAB	4.004
		14640	Millout Extension		4.343
			Tubing (15 JTS) 4–1/2" 12.6 PPF	VAM L80 9 CR	3.968
14642-14657		14721	Sliding Sleeve (Closed)	Otis XD	3.313
14661-14697					
14714-14724					
14729-14733		14736	Anchor	Baker K-22	3.800
		14740	Permanent Packer	Baker FAB	4.004
		14743	Millout Extension		4.353
		14750	Landing Nipple	Otis XN	2.635
		14750	Nipple Plug		
		14761	Muleshoe		3.958
14756-14785					
14700-14811					
14816-14819					
14816-14828					
14933-14838					
14856-14864					
14901-14899		14999	Top of Cement		
		15006	Cement Retainer		
15019-15116		15141	Cement Retainer		
		15225	TD		

Figure 12-27 Example of Ninian Selective Completion; MD Measured Depth; TD Total Depth; and ID Inner Diameter (Copyright 1996, SPE, from JPT, December 1996[12])

- Utilize chrome tubing to reduce failure due to corrosion
- Increase life of well by chrome tubing coupled with the installation of selective completions
- Use successful scale inhibition program and the use of monobase completions

Production Forecasts

The future performance of the Ninian waterflood has been evaluated using a number of techniques. These are given in Table 12-4 and indicate the ultimate recovery between 1,120 and 1,250 MMBO.

Table 12-4 Production Forecasts (Copyright 1996, SPE, from JPT, December 1996[12])

	Ultimate Recovery MMBO
Exponential decline	1,120
Log water-oil ratio vs cumulative oil recovery	1,256
X-plot	1,256
NP (after water injection) vs NP/cumulative water injection	1,256
History-matched reservoir simulation	1,178
(Economic limit = 97% water cut or 15,000 BOPD)	

Reservoir Management Issues/Challenges

Although Ninian is a mature waterflood, there may exist following opportunities to improve recovery:

- Continually evaluate the performance to achieve optimum alignment between injectors and producers
- Investigate bypassed oil opportunities
- Identify well-work program based upon zonal contribution investigation
- Develop attic oil between current producers and faults/erosional edges updip

References

1. Wayhan, D. A. and J. A. McCaleb. "Elk Basin Madison Heterogeneity —
    Its Influence on Performance," *JPT* (February 1969) 153-159.
2. Ghauri, W.K. "Innovative Engineering Boosts Wasson Denver Unit
    Reserves," Petroleum Engineer (December 1974) 26-34.
3. Ghauri, W.K. "Waterflood Surveillance," Paper Presented at the 23rd
    Annual Southwestern Petroleum Short Course, (1976), 195-199.
4. Ghauri, W.K. "Production Technology Experience in a Large Carbonate
    Waterflood, Denver Unit, Wasson San Andres Field," *JPT* (September
    1980), 1493-1502.
5. Ghauri, W.K., A. F. Osborne, and W. L. Magnuson. "Changing Concepts on
    Carbonate Waterflooding - West Texas Denver Unit Project - An
    Illustrative Example," *JPT* (June 1974), 596-606.
6. Hunter, J.D. et al. "Denver Unit Well Surveillance and Pump-off Control
    System," *JPT* (September 1978), 1319-1326.
7. Stiles, L.H. and J. B. Magruder. "Reservoir Management in the Means San
    Andres Unit," *JPT* (April 1992) 469-475.
8. Langston, E.P. "Field Application of Pressure Buildup Tests, Jay-Little
    Escambia Fields," SPE Paper 6199, presented at the 1976 SPE Annual
    Technical Conference and Exhibition, October 3-6, New Orleans, LA.
9. Langston, E. P. and J. A. Shirer. "Performance of Jay/LEC Fields Unit
    Under Mature Waterflood and Early Tertiary Operations," SPE Paper
    11986, presented at the 1983 SPE Annual Technical Conference and
    Exhibition, October 5-8, San Francisco, CA.
10. Langston, E. P., J. A. Shirer, and D. E. Nelson. "Innovative Reservoir
    Management — a Key to Highly Successful Jay/LEC Waterflood," SPE
    Paper 9476, presented at the 1980 SPE Annual Technical Conference and
    Exhibition, September 21-24, Dallas, TX.
11. Shirer, J. A. "Jay/LEC Waterflood Pattern Performs Successfully," SPE
    Paper 5534, presented at the 1975 SPE Annual Technical Conference and
    Exhibition, September 28-October 1, Dallas, TX.
12. Omoregie, Z. S. et al. "Reservoir Management in the Ninian Field — a
    Case History," *JPT* (December 1996) 1136-1138.
13. Willhite, G. P. "Waterflooding," SPE Textbook Series Vol. 3 (1986).

13

Current and Future Challenges

The current challenges primarily focus on (A) improved reservoir definition, (B) tracking the movement of injected water vertically and areally to maximize the vertical and areal sweep efficiencies, and (C) to control that movement. The problem boils down to contacting the maximum possible volume of the reservoir—any technology development to achieve this will be of great value in waterflood asset management.

Outlook and the Next Step

In the years ahead, more attention will be given to managing waterflood effectively from day one. The areas that will play an increasing role are:

- *Improved Reservoir Definition*

A detailed description of the reservoir involving spatial variation and continuity of porosity, permeability, and fluid saturation is required. An integrated geoscience-engineering model provides information about the likely fluid flow-paths and an opportunity to improve the design and performance of a waterflood.

More detailed reservoir descriptions involving geostatistics, cross-well tomography and 3D seismic are NOW being obtained. It is anticipated that geophysical techniques, combined with other geological and reservoir engineering technologies, will play important roles in describing the reservoir and in the surveillance of an ongoing waterflood. Four-dimensional seismic technology, coupled with logging and coning of observation wells, tracer monitoring, and reservoir simulation of waterflooding process will play key roles in the future.

- *Tracking the movement of injected water vertically and areally to maximize the vertical and areal sweep efficiencies*

As discussed before, the maximization of areal and vertical sweeps lead to maximum oil recovery in a waterflood. Also, there is a profound effect of the proper timing of the implementation of volumetric sweep technology on recovery efficiency. An increase from 20 to 45% is possible in some cases. Therefore, it is important to maximize sweep early in the project's life. Trying to squeeze more oil from marginal projects will, at best, yield marginal to average results. Designing waterfloods with

- dedicated observation/monitoring wells
- 3D and 4D seismic technology applications to place injectors and producers appropriately
- use of horizontal well technology to drain bypassed oil
- application of tracer technology
- implementing volumetric sweep improvement technologies will become more popular for managing waterfloods effectively

- *Control the movement of injected water*

Waterflood surveillance techniques, documented in Chapter 9, play an important role in monitoring the movement of injected water. Controlling the movement of injected water can assist in maximizing the volumetric sweep. Vertical sweep improvements by mechanical or chemical means are absolutely necessary. Areal flood balancing, appropriate placement (number and location) of injection/production wells, and tapping the bypassed oil are various means of improving areal sweep efficiency.

Timing of implementation of control methods for the movement of injected water is critical, because a very significant difference in recovery efficiency exists between a well-managed and a poorly-managed waterflood. As much as a 25% difference in OOIP recovery efficiency is possible in some cases.

- *Field automation and IMS (Information Management System)*

Field automation will continue to play an increasingly important role in waterflood asset management. Automation coupled with IMS will enhance technical analysis and control. Our skills in controlling future performance will depend on how well we document and utilize the past and the present.

Field automation makes data collection easier and subsequently any changes to improve the management of the existing waterflood—e.g., adjusting pressure and rate of any injection well—is easily achieved. As a result, pattern balancing becomes a routine operation in the field.

Development and application of integrated computer software using artificial intelligence (an expert system) will be of great importance in the future. In addition, it is easier to transfer expertise to new professionals if a well-organized IMS is available. The transfer of knowledge in a team-based, multifunctional organization is becoming a major challenge, and an IMS can be a good tool toward achieving this goal.

- *Reservoir/asset management—a team approach*

A systems approach that includes reservoir characterization, fluids and their behavior in the reservoir, creation and operation of wells, and surface processing of fluids is needed. Economic recovery from a waterflood can be maximized by an integrated group effort. All decisions pertaining to the waterflood should be made by the team considering the entire system, rather than only one aspect. The team approach to waterflood asset management should always be emphasized.

Appendix

A

Reservoir Management Concepts and Process

The need to enhance recovery from the vast amount of remaining oil and gas-in-place in the United States and elsewhere in the world, and the global competition requires better reservoir management practices. Historically, some form of reservoir management has been practiced when a major expenditure is planned, such as a new field development or waterflood installation. The reservoir management studies at these instances were not integrated, i.e., different disciplines did their own things separately. During the last 20 years, however, greater emphasis has been put on synergism between engineering and geosciences. Despite the emphasis, progress on integration has been slow.

A reservoir's life begins with exploration leading to discovery followed by delineation of the reservoir, development of the field, production by primary, secondary and tertiary means, and finally to abandonment (Figure A-1).[1,2] Integrated sound reservoir management is the key to successful operation throughout a reservoir's entire life.

This appendix provides:

1. Reservoir management definition
2. Reservoir management process
3. Synergy and team work
4. Organization and team management
5. Integration of geoscience and engineering
6. Reasons for failure of reservoir management program

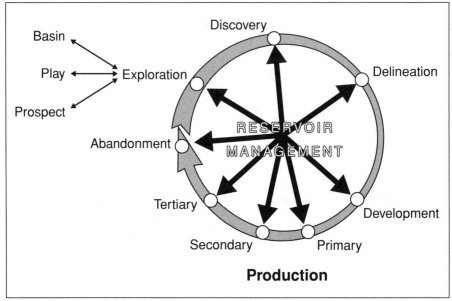

Figure A-1 Reservoir Life Process (Copyright 1992, SPE, from paper 22350)

Reservoir Management Definition

The goal or definition of reservoir management can be varied depending upon the political and/or economic environment in which it is applied: conserving energy, maximizing recovery, maximizing profitability, and even maximizing employment. Considering economic environment, a number of authors have defined reservoir management recently.[2-6] Sound reservoir management requires judicious use of various means available to maximize benefits from a reservoir. It relies on utilizing human, technological and financial resources to maximize profits by optimizing recovery while minimizing capital investments and operating expenses (see Figure A-2).

Reservoir Management Process

The modern reservoir management process[2] involves (see Figure A-3):

1. Setting purpose or strategy
2. Developing economically viable plan

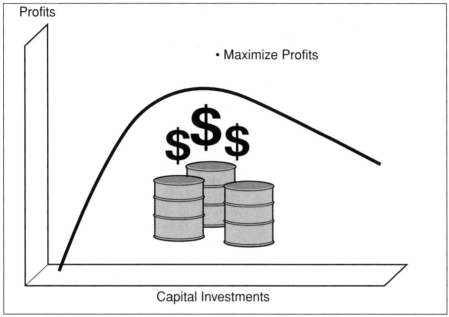

Figure A-2 What Is Reservoir Management? (Copyright 1992, SPE, from paper 22350)

3. Implementing plan
4. Monitoring performance
5. Evaluating performance
6. Revising plans and strategies

The components of the reservoir management process are interdependent and their integration is essential for successful reservoir management. The management process is dynamic, and ongoing. Monitoring, surveillance, evaluation and revision are often neglected today. As additional data become available, the management process needs refinement and implementation with appropriate changes.

The success in reservoir management requires goal setting, deliberate planning, implementing, on-going monitoring and evaluating the reservoir performance, keeping short term but more importantly long term goals. While a comprehensive plan for reservoir management is highly desirable, every reservoir may not warrant such a detailed plan because of cost effectiveness. However, the key to success is to have a management plan (whether so comprehensive or not) and implement it from day one.

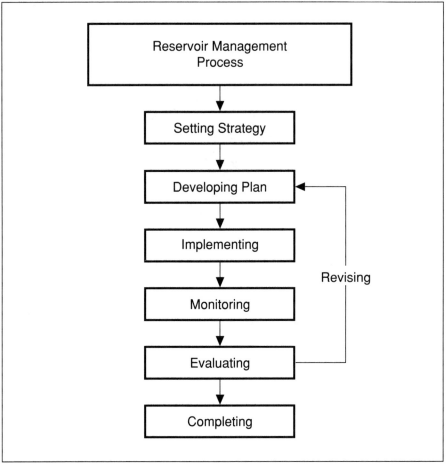

Figure A-3 Reservoir Management Process (Copyright 1992, SPE, from paper 22350)

Setting Purpose or Strategy

Recognizing the specific need, and setting a realistic and achievable purpose is the first step in reservoir management. The key elements for setting reservoir management goals are:

1. Reservoir characteristics
2. Business environment
3. Political situation
4. Available technology

Understanding the nature of the reservoir requires a knowledge of the geology, rock and fluid properties, fluid flow and recovery mechanisms, drilling and well completions and past production/injection performance (see Figure A-4).

Understanding corporate goal, financial strength, culture and attitude, business climate, oil/gas price, inflation, capital and personnel availability, conservation, safety and environmental regulations, and political stability is of vital importance.

Utilizing fully available technology in exploration, drilling and completions, recovery processes, and production ensures success in reservoir management. Many technological advances have been made in all of these areas.

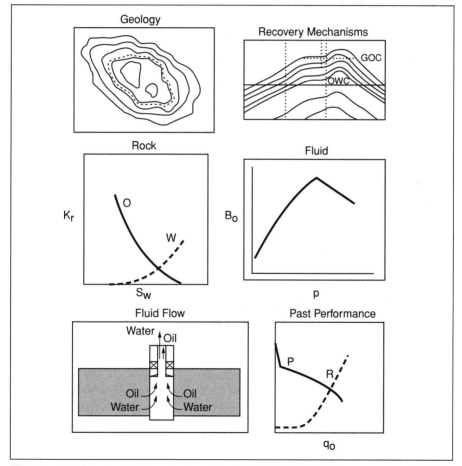

Figure A-4 Reservoir Knowledge (Copyright 1992, SPE, from paper 22350)

Developing Economically Viable Plan

A comprehensive reservoir management plan is essential for the success of a project. It involves many time-consuming development steps:

1. Development and depletion strategy
2. Environmental considerations
3. Data acquisition and analysis
4. Reservoir model building
5. Production and reserves forecasts
6. Facilities requirements
7. Economic optimization
8. Management approval

Development and depletion strategies will depend upon the stage at which the reservoir is going through. In case of a new discovery, we need to address the question as to the best way to develop the field; well spacing, number of wells, recovery schemes, i.e., primary and subsequently secondary and tertiary. If the reservoir has been depleted by primary means, secondary and even tertiary recovery schemes need to be investigated.

In developing and subsequently operating a field, environmental and ecological considerations have to be included; regulatory agency constraints will have to be also satisfied. These are very sensitive and important issues of the reservoir management process.

Reservoir management requires a thorough knowledge of the reservoir which is gained through an integrated data acquisition and analysis program involving all functions from the beginning. An efficient data management program consisting of collection, analyses, storing and retrieving lays the foundation of sound reservoir management. The program should establish the need for the data with a cost/benefit analysis.

An integrated geoscience and engineering reservoir model needs to be built jointly by geoscientists and engineers. The geological model, particularly the definition of geological units and their continuity and compartmentalization, is an integral part of geostatistical and ultimately reservoir simulation models. The engineering model is concerned with rock and fluid properties, fluid flow and recovery mechanisms, drilling, completions, production, and injection.

Evaluation of the past and present reservoir performance and forecasts of its future behavior under various scenarios is an essential aspect of reservoir management process (see Figure A-5). The accuracy

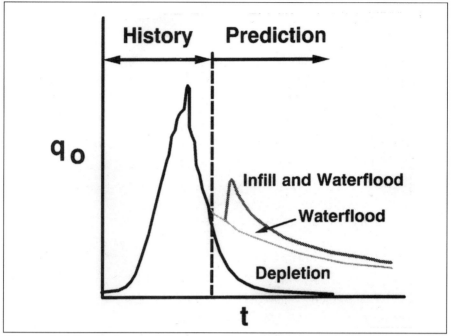

Figure A-5 Production and Reserves Forecasts (Copyright 1992, SPE, from paper 22350)

of the reservoir performance analysis is dictated by the quality of the reservoir model. Techniques used for evaluation of the past and present reservoir performance and forecast of its future behavior are classical volumetric, material balance and decline curve analysis methods, and high-technology black oil, compositional and enhanced oil recovery numerical simulators.

Production performance results are used to estimate facilities requirements. Facilities are the actual physical link to the reservoir. Everything we do to the reservoir, we do through the facilities. These include drilling, completion, pumping, injecting, processing, and storing. Proper design and maintenance of facilities has a profound effect on profitability. The facilities must be capable of carrying out the reservoir management plan but they cannot be wastefully designed. Estimates of the capital and operating costs based on the facilities requirements are used for economic analyses.

With estimated production, capital, operating expenses, and financial data, project economics are evaluated. Economic optimization is the ultimate goal of reservoir management.

The final step in developing a reservoir management plan requires management approval and support.

Implementing Plan

After management approval of the project development plan, the next major assignment is to implement it to get production on stream as soon as possible. In case of developing a field for primary or waterflood recovery, the various activities are as follows:

1. Design, fabricate, and install surface and subsurface facilities
2. Develop a drilling/completion program
3. Acquire and analyze necessary logging, coring, and well test data
4. Upgrade the reservoir database and revise production and reserves forecasts

The key steps for successfully implementing a plan are:

1. Start with a plan of action involving all functions
2. Make plan flexible
3. Have full management support with involvement from "day one"
4. Have support and commitment of field personnel
5. Have periodic review meetings, involving all team members, mostly in the field offices

The important reasons for failure to successfully implement a plan are:

1. Lack of overall knowledge of the project on the part of all team members
2. Failure to interact and coordinate the various functional groups
3. Delay in initiating the management process

Monitoring Performance

Sound reservoir management requires constant monitoring and surveillance of the reservoir performance as a whole in order to determine if the reservoir performance is conforming to the management plan. In order to successfully carry out the monitoring and surveillance program, coordinated efforts of the various functional groups working on the project are needed at the start of the project.

An integrated and comprehensive program needs to be developed for successful monitoring and surveillance of the management project. The engineers, geologists, and operations personnel should work together on the program with management support and field personnel commitment. Dedicated and coordinated efforts of the various functional groups are essential.

The program will depend upon the nature of the project. Ordinarily, the major areas of monitoring and surveillance involving data acquisition and management include:

1. Oil, water and gas production
2. Gas and water injection
3. Systematic and periodic static and flowing bottom hole pressure testing of selected wells
4. Production and injection tests
5. Production and injection profiles
6. Recording of workovers and results
7. Any others aiding surveillance

Evaluating Performance

The management plan must be reviewed periodically to ensure that it is being followed, that it is working, and that it is still the best plan. The success of the plan needs to be evaluated by checking the actual reservoir performance against the anticipated behavior.

It would be unrealistic to expect the actual project performance to match exactly the planned behavior. Therefore, certain technical and economic criteria need to be established by the functional groups working on the project to determine the success of the project. The criteria will depend upon the nature of the project. A project may be a technical success but an economic failure.

How is the reservoir management plan working? The answer lies in the careful evaluation of the project performance. The actual performance, for example, reservoir pressure, gas-oil-ratio, water-oil-ratio, production, and injection needs to be routinely compared with the expected (see Figure A-6). In the final analysis the economic yardsticks will determine the success or failure of the project.

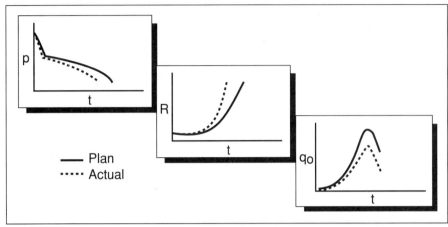

Figure A-6 Evaluation (Copyright 1992, SPE, from paper 22350)

Revising Plans and Strategies

Revision of plans and strategies is needed when the reservoir performance does not conform to the management plan or when conditions change. The answers to such questions as is it working, what needs to be done to make it work, what would work better, etc. must be asked and answered on an ongoing basis in order for us to say we are practicing sound reservoir management.

Synergy and Team Work

Halbouty, Sneider, Talash, Harris and Hewitt advocated synergy and team efforts for successful reservoir management. Synergy means that geologists, geophysicists, petroleum engineers and others work together on a project more effectively and efficiently as a team than working as a group of individuals. The team approach to reservoir management is essential involving interaction between management, engineering, geoscience, research and service functions. The synergism of the team approach can yield a "whole is greater than the sum of its parts" effect.[7, 8, 9, 10]

It is recognized more and more that reservoir management is not synonymous with reservoir engineering and/or reservoir geology. Success requires multidisciplinary, integrated team efforts. The players are everybody who has anything to do with the reservoir (Figure A-7).[2] The team members must work together to ensure development and execution of the management plan. By crossing the traditional boundaries and integrating their functions, corporate resources are better utilized to achieve the common goal.

Nowadays, it is becoming common for large reservoir studies to be integrated through a team approach. However, mere creation of a team does not guarantee integration leading to success. Team skills, team authority, team compatibility with the line management structure, and overall understanding of the reservoir management process by all team members are essential for the success of the project. Also, most reservoir management teams are being assembled only at key investment times. What is missing today is ongoing multidisciplinary reservoir management efforts for all significant reservoirs.

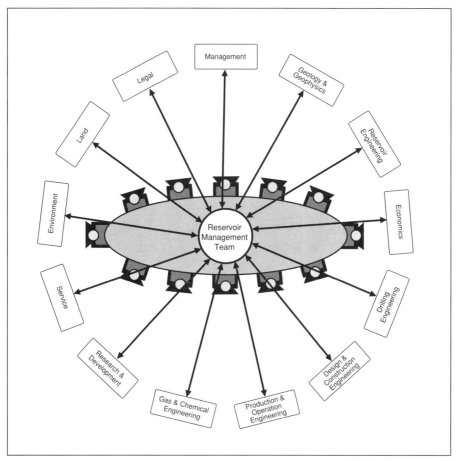

Figure A-7 Reservoir Management Team (Copyright 1992, SPE, from paper 22350)

A team approach to reservoir management can be enhanced by the following:

1. Facilitating communication among various engineering disciplines, geology and operations staff by: (a) meeting periodically, (b) interdisciplinary cooperation in teaching each other's functional objectives, and (c) building trust and mutual respect. Also, each member of the team should learn to be a good teacher

2. The engineer, to some degree, developing the geologist's knowledge of rock characteristics and depositional environment, and the geologist cultivating knowledge in well completion and other engineering tasks, as they relate to the project at hand

3. Each team member subordinating his ambitions and egos to the goals of the reservoir management team, and maintaining a high level of technical competence

4. The team members working as a well-coordinated "basketball team" rather than a "relay team"

Organization and Team Management

In addition to synergy and team approach, the organization and management of the team are also very critical. Recognizing that sound reservoir management requires a multidisciplinary team effort, the important question then is, who should set goals and make reservoir management decisions. Should the production manager make the decisions or should they be made by the team?

In the old or the conventional system (see Figure A-8), various members of the team (i.e., geologist, reservoir/production/facilities engineers, operations staff, and others) work on a reservoir under their own bosses/functional heads. In the new multidisciplinary team approach, the team members from various functions work on a given reservoir under a team leader and sometimes operate as a self-managed team (see Figure A-9). The team leader generally provides day-to-day guidance, and the so-called functional heads or gurus give occasional functional guidance. The team (or its members) do not administratively report to the functional gurus. Administrative and project guidances are provided by the production manager or the asset manager. The asset management concept emphasizes focusing on a field as an asset and all team members have a primary objective of maximizing the short- and

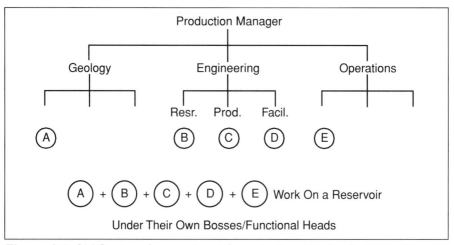

Figure A-8 Old System-Conventional Organization (Abdus Satter and Ganesh Thakur, Integrated Petroleum Reservoir Management: A Team Approach, Copyright PennWell Books, 1994)

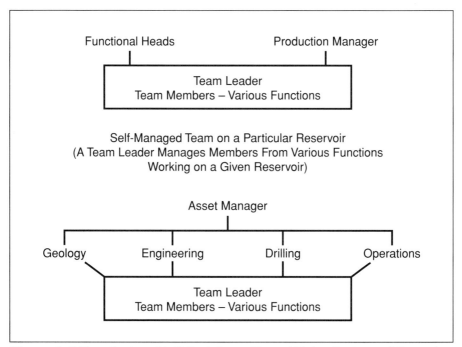

Figure A-9 New System-Multidisciplinary Team (Abdus Satter and Ganesh Thakur, Integrated Petroleum Reservoir Management: A Team Approach, Copyright PennWell Books, 1994)

long-term profitability of the asset. The team members concentrate on their duties more like "generalists" than functional specialists.

In order to make effective decisions, the production manager has to recognize the dependence of the entire system upon the nature and behavior of the reservoir. However, it is rare to have someone with expertise in all required areas. In our experience, we have observed that reservoir management plan development and implementation are most effective if the team members work together and are involved in decision making.

Organization and management of the reservoir management team requires special attention. Design of the team, selection of team members, appropriate motivational tools, and composition of the team (as the needs of the reservoir change) should be carefully considered. Other aspects, such as team leadership, establishment of team goals and objectives, and performance appraisals of the team members are some matters that play a key role in effective reservoir management.

Once a team is designed and begins to function, another item of significance is how to sustain team effort. It is easy to have excitement when the teams are set up, at times of major expenditures, development effort, 3-D seismic program, etc.; however, to have ongoing attention by a multidisciplinary team for all major reservoirs requires great commitment by the operating company.

One model of the team management is illustrated below:

1. Functional management nominates team members to work on a project with specific tasks in mind
2. The team members consist of representatives from geology and geophysics, various engineering functions, field operations, drilling, finance, etc
3. The team selects a team leader, whose responsibility is to coordinate all activities and report to the production manager
4. Team members prepare a reservoir management plan and define their goals and objectives by involving all functional groups. The plan is then presented to the production manager, and after receiving his feedback, appropriate changes are made. Next the plan is published and all members follow the plan
5. The team members' performance evaluation is conducted by their functional heads with input from the team leader and the production manager. The performance appraisal, in addition to various dimensions of his/her performance, includes teamwork as a job requirement

6. Teams are rewarded recognition/cash awards upon timely and effective completion of their tasks. These awards provide an extra motivation for team members to do well
7. As the project goals change, e.g., from primary development to secondary process, the team composition changes to include members with the required expertise. Also, this provides an opportunity to change/rotate team members with time
8. Approvals for project AFE's are initiated by the team members; however, the engineering/operations supervisor and/or production manager have the final approval authority
9. Sometimes conflicting priorities for the team members develop because they essentially have two bosses, i.e., their functional heads and the team leader. These conflicts are generally resolved by constant communication between the team leader, functional heads, and the production manager

Integration Of Geoscience and Engineering

Synergy and team concepts are the essential elements for integration of geoscience and engineering. Various authors[11-14] advocated the need for increased interaction, coordination, and integrated cross-training between geoscientists and engineers to advance petroleum exploration, development, and production. In fact, integration of people, technology, tools, and data is vital for the success of a project[2] (see Figure A-10).

Success for integration depends on:

1. Overall understanding of the reservoir management process, technology, and tools, through integrated training and integrated job assignments
2. Openness, flexibility, communication, and coordination
3. Working as a team
4. Persistence

Integration of geological, geophysical, and engineering data is essential to develop reservoir description model which is used to analyze reservoir performance.[13] Both engineering and geological judgment must guide the development and use of the simulation model. It is

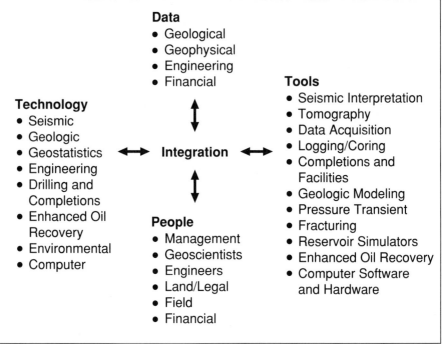

Data
- Geological
- Geophysical
- Engineering
- Financial

Technology
- Seismic
- Geologic
- Geostatistics
- Engineering
- Drilling and Completions
- Enhanced Oil Recovery
- Environmental
- Computer

Integration

People
- Management
- Geoscientists
- Engineers
- Land/Legal
- Field
- Financial

Tools
- Seismic Interpretation
- Tomography
- Data Acquisition
- Logging/Coring
- Completions and Facilities
- Geologic Modeling
- Pressure Transient
- Fracturing
- Reservoir Simulators
- Enhanced Oil Recovery
- Computer Software and Hardware

Figure A-10 Reservoir Management (Harris, J.G., The Role of Geology in Reservoir Simulation Studies, *JPT*, May, 1975)

important that the geologist and the engineer understand each other's data. Throughout his work, the geologist requires input and feedback from the engineer. Examples of this "interplay of effort" are indicated in Figure A-11.[13]

Reservoir engineers and geologists are beginning to benefit from seismic and cross-hole seismology data. Also, it is essential that geological and engineering ideas and reasoning be incorporated into all seismic results if the full economic value of the seismic data is to be realized. Geological and engineering data should be reviewed and coordinated with the geophysicists to determine whether or not an extension is possible for the drilling of an exploratory well.

The application of 3D-seismic analysis plays an important role in reservoir management. A 3D-seismic analysis can lead to identification of reserves that may not be produced optimally, or perhaps not produced at all, by the existing reservoir management plan. In addition, it can save costs by minimizing dry holes and poor producers.

The initial interpretation of a 3D-seismic survey affects the original development plan. With the development of the field, additional

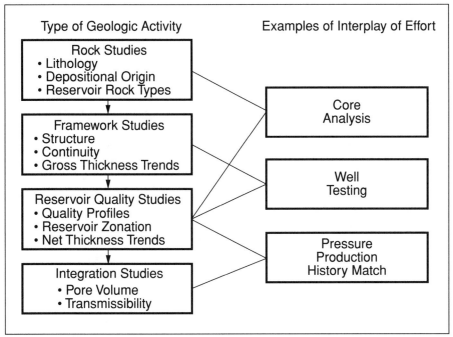

Figure A-11 General Geological Activities in Reservoir Description and Input From Engineering Studies (Copyright 1992, SPE, from paper 22350)

information is collected and is used to revise and refine the original interpretation. The usefulness of a 3D-seismic survey lasts for the life of the reservoir.

Woofer, and MacGillivary reported a very successful development plan of the Brassey oil field by an aggressive engineering/geoscience team.[14] The team approach combined the knowledge from a geologic model, seismically defined pool edges, continuity information from well-test data, and material balance calculations to predict reservoir volume, areal extent, and continuity on the basis of an integrated reservoir model.

Reasons For Failure of Reservoir Management Program

There are numerous reasons why reservoir management programs have failed. Some are listed next:

Unintegrated System

Management program was not considered as a part of a coupled system consisting of wells, surface facilities, and the reservoir. Not all of these were emphasized in a balanced way. For example, one could do well in studying the fluids and their interaction with rock, i.e., reservoir engineering; but, by not considering the well and/or the surface system design, the recovery of oil and/or gas was not optimized.

Perhaps the most important reason why a reservoir management program is developed and implemented poorly is unintegrated group effort. Sometimes the operating decisions are made by people who do not recognize the dependence of one system on the other. Also, the people may not have the required background knowledge in critical areas, e.g., reservoir engineering, geology and geophysics, production and drilling engineering, and surface facilities.

Starting Too Late

Ideally, reservoir management should start on day one of a project. Often, reservoir management is not started early enough and when initiated, the management became necessary because of a crisis that occurred and/or required a major problem to be fixed. Early initiation of a coordinated reservoir management program could have provided a better monitoring and evaluation tool, and have cost less in the long run. For example, a few early DSTs, could have helped decide if and where to set pipe. Performing some tests early on could have indicated the size of the reservoir. If it were of limited size, drilling of unnecessary wells could have been prevented. Also, injection and production profiles in waterflooding projects would have shown fluid entry and production intervals indicating effectiveness of the recovery projects.

Lack of Maintenance

Calhoun draws an analogy between reservoir and health management.[15] According to his concept, it is not sufficient for the reservoir management team to determine the state of a reservoir's health and then attempt to improve it. One reason for reservoir management ineffectiveness is that the reservoir and its attached system's (wells and surface facilities) health (condition) is not maintained from the start.

References

1. Satter, A. S., and Thakur, G. C. "Integrated Petroleum Reservoir Management: A Team Approach," *PennWell Books*, Tulsa, OK (1994) .
2. Satter, A., J. E. Varnon, and M. T. Hoang. "Integrated Reservoir Management," JPT (Dec 1994)
3. Thakur, G. C. "Reservoir Management: A Synergistic Approach," paper SPE20138 presented at the 1990 SPE Permian Basin Oil and Gas Recovery Conference, Midland, TX March 8-9.
4. Haldorsen, H. H. and T. Van Golf-Racht. "Reservoir Management Into the Next Century," Paper NMT 890023 presented at the Centennial Symposium at New Mexico Tech., Socorro, NM, Oct. 16-19, 1989.
5. Wiggins, M. L. and R. A. Startzman. "An Approach to Reservoir Management," SPE Paper 20747, Reservoir Management Panel Discussion, SPE 65th Ann. Tech. Conf. & Exhib., Sept. 23-26, 1990, New Orleans, LA.
6. Robertson, J. D. "Reservoir Management Using 3D Seismic Data," *JPT* (July 1989).
7. Halbouty, M. T. "Synergy Is Essential to Maximum Recovery,"*JPT* (July 1977), pp.750-754.
8. Sneider, R. M. "The Economic Value of a Synergistic Organization," Archie Conference, Oct. 22-25, 1990 Houston, TX.
9. Talash, A. W. "An Overview of Waterflood Surveillance and Monitoring," *JPT* (Dec. 1988) 1539-43.
10. Harris, D. G. and C. H. Hewitt. "Synergism in Reservoir Management - The Geologic Perspective," *JPT* (July 1977) pp. 761-770.
11. Satter, A. "Reservoir Management Training - An Integrated Approach," SPE paper 20752, Reservoir Management Panel Discussion, SPE 65th Annual Technical Conference & Exhibition, Sept. 23-26, 1990, New Orleans, LA.
12. Robertson, J. D. "Reservoir Management Using 3-D Seismic Data," *Geophysics: The Leading Edge of Exploration* (Feb. 1989) 25-31.
13. Harris, D. G. "The Role of Geology in Reservoir Simulation Studies," *JPT* (May 1975) 625-32.
14. Woofer, D. M., and MacGillivary. "Brassey Oil Field, British Columbia: Development of an Aeolian Sand - A Team Approach," SPE Reservoir Engineering (May 1992) 165-172.
15. Calhoun, J. C. "A Definition of Petroleum Engineering," *JPT* (July 1963) 725-727.

APPENDIX
B

Reservoir Model

This appendix summarizes an approach to building an integrated reservoir model based upon geological, geophysical, petrophysical, and engineering data, as described in reference.[1]

The reservoir model is not just an engineering or a geoscience model; rather it is an integrated model, prepared jointly by geoscientists and engineers. The integrated reservoir model requires a thorough knowledge of the geology, rock and fluid properties, fluid flow and recovery mechanisms, drilling and well completions, and past production performance.

The geological model is derived by extending localized core and log measurements to the full reservoir using many technologies such as geophysics, mineralogy, depositional environment, and diagenesis. The geological model (particularly the definition of geological units and their continuity and compartmentalization) is an integral part of geostatistical and ultimately reservoir simulation models.

Role of Reservoir Model

The economic viability of a petroleum recovery project is greatly influenced by the reservoir production performance under the current and future operating conditions. Therefore, the evaluation of the past and present reservoir performance and the forecast of its future behavior is an essential aspect of the reservoir management process. Classical volumetric, material balance and decline curve analysis methods, and high-technology, black oil, compositional and enhanced oil recovery numerical simulators are used for analyzing reservoir performance and estimating reserves. The accuracy of the results is dictated by the quality of the reservoir model used to make reservoir performance analysis.

As opposed to one reservoir life, the simulator can simulate many lives for the reservoir under different scenarios and thus provide a very powerful tool to optimize the reservoir operation. Thomas discusses with examples the role reservoir simulators play in formulating initial development plans, history matching and optimizing future production, and in planning, and designing enhanced oil recovery projects.[2] Figure B-1 presents key steps involved in reservoir modeling and performance prediction.[3,4]

Historically, reservoir simulators have been used for studying large fields and those undergoing complex recovery processes. They have not been suitable for small reservoirs because the simulation studies are costly, require highly trained professionals, and are too time consuming for an operating company work environment. This is changing. A proposed minisimulation technique using personal computers can

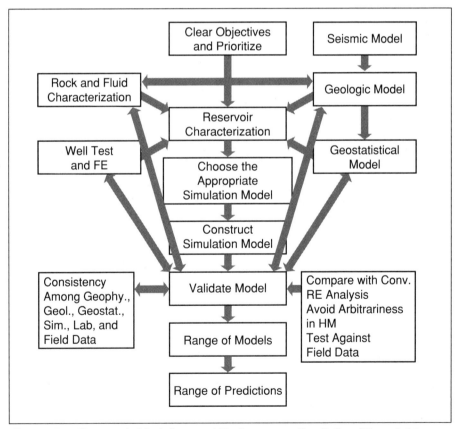

Figure B-1 Reservoir Modeling and Performance Prediction (Copyright 1996, SPE, from JPT, June 1996[3, 4])

play an important role in managing small reservoirs.[5]

The process of developing a sound reservoir model plays a very important part in reservoir management because:

- It requires integration among geoscientists and engineers
- It allows geoscientists' interpretations and assumptions to be compared to actual reservoir performance as documented by production history and pressure tests
- It provides a means of understanding the current performance and predicts the future performance of a reservoir under various "what if" conditions so that better reservoir management decisions can be made[6]

In addition, the reservoir model should be developed jointly by geoscientists and engineers because:

- An interplay of effort results in better description of the reservoir and minimizes the uncertainties of a model. The geoscientists' data assist in engineering interpretations, whereas the engineering data sheds new light on geoscientists' assumptions
- The geoscientist-engineer team can correct contradictions as they arise, preventing costly errors later in the field's life
- It has been well documented cross-functional team can gain valuable insight in describing a reservoir. For example, while working in a complex fault system, Amoco International Oil Company's reservoir engineers worked with geologists to "produce an accurate *a priori* reservoir description." The team tested the description against field performance using a simulation model and gained invaluable information related to fault configuration and the relationships of gas in place, permeability, and reserves. The model led to confident planning of future platform and compression requirements, and it provided much needed lead time to install equipment[26]
- In a fragmented effort (i.e., when engineers and geoscientists do not communicate), each discipline may study only a fraction of the available data; thus, the quality of the reservoir management can suffer and adversely affect drilling decisions and depletion plans throughout the life of the reservoir
- Multidisciplinary teams using the latest technology provide opportunities to tap unidentified reserves. For example, improved 3-D seismic data can aid the surveillance of production operations in mature projects and can identify the

> presence or lack of continuity between wells, and thus improve the description of the reservoir model

- Utilizing reservoir models developed by multidisciplinary teams can provide practical techniques of accurate field descriptions to achieve optimal production

Thus, it is important that we prepare a simulation model that takes into account realistic geology and other rock-fluid characteristics. With a realistic simulation model, we can do or obtain guidance on the following:

- Determine the performance of an oil field under water injection or gas injection, or under natural depletion
- Judge the advisability of flank waterflooding as opposed to pattern waterflooding
- Determine the effects of well location and spacing
- Estimate the effect of producing rate on recovery
- Calculate the total gas field deliverability for a given number of wells at certain specified locations
- Estimate the lease-line drainage in heterogeneous oil or gas fields[6,7]

Breitenbach presents a comprehensive history of reservoir simulation, starting in the 1940s.[8] In the 1940s analog models played an important role. During the 1950s, with the advent of 2D and 3-D finite difference equations, it became possible to solve problems related to multiphase flow in heterogeneous porous media. In the 1960s, reservoir simulation was devoted largely to three-phase, black oil reservoir problems. During the 1970s, the efforts extended to enhanced oil recovery processes. In the 1980s, the reservoir simulation application continued to expand and included the area of reservoir description. The use of geostatistics to describe reservoir heterogeneities and technology to model naturally fractured reservoirs were developed.[9-11]

Today, desktop computers and a wide variety of reservoir simulation systems provide engineers and geoscientists an economical means to solve complex reservoir problems in a reasonable time frame.

The simulation process consists of describing the reservoir (i.e., preparing a reservoir model), matching historical performance, and predicting the future performance of the reservoir under a variety of scenarios. The reservoir description step results in a reservoir model that includes overall geometry (i.e., permeability, porosity, and height of each grid block).[12-14] After constructing the reservoir model, it is generally validated by determining whether it can duplicate past field behavior.

Often the reservoir description information is changed within geological and engineering bounds to history-match the past production performance.

There are many educational values of simulation models:

- Too often we tend to demand accurate determination of all types of input data before we accept the computed results as meaningful or reliable. On the other hand, interest in accuracy of input data should be proportional to the sensitivity of computed results to variations in those data
- Sensitivity to errors in reservoir description data can be determined by performing simulation runs with variations in those data covering a reasonable range of uncertainty[6,7,15]

Thus, efforts should be made on obtaining those data that have the greatest effect on calculated performance. Also, a general guide for developing a model is to "select the least complicated model and most gross reservoir description that will allow the desired estimation of reservoir performance."[7]

Many authors have documented applications of 3-D seismic in reservoir management and modeling.[16,17] Geophysicists are playing a more important role than ever in identifying key reservoir features in a simulation model. Also, they are assisting in validating the geologic model.

Geoscience

Geoscientists probably play the most important role in developing a reservoir model. The purpose of this section is to provide the geoscientist's activities in developing a reservoir model. The model requires variations in porosity, permeability, and capillary properties.

The distributions of the reservoir and nonreservoir rock types and of the reservoir fluids determine the geometry of the model and influence the type of model to be used.[18] For example, the number and scale of the shale (or dense carbonate) breaks in the physical framework determine the continuity of the reservoir faces and influence the vertical and horizontal dimensions of each cell. Real variations in reservoir parameters may require several cross-sections or a three-dimensional model. Other influences on cell dimension include computing cost, well spacing, fluid-phase distribution, and the purpose of the study.

Incorporation of a geologic model into a simulation model requires recognition and capture of detailed reservoir heterogeneities.

With the advent of advanced simulation technology and the understanding of complex subsurface structures, these heterogeneities can be recognized early in the life of a field and incorporated into the simulation model.

Both engineering and geological judgment must guide the development and use of the simulation model. The geologist usually concentrates on the rock attributes in four stages:

1. Rock studies establish lithology and determine depositional environment, and reservoir rock is distinguished from non-reservoir rock

2. Framework studies establish the structural style and determine the three-dimensional continuity character and gross-thickness trends of the reservoir rock

3. Reservoir-quality studies determine the framework variability of the reservoir rock in terms of porosity, permeability, and capillary properties (the aquifer surrounding the field is similarly studied)

4. Integration studies develop the hydrocarbon pore volume and fluid transmissibility patterns in three dimensions[18,19]

Geoscientists and engineers need feedback from each other throughout their work, and an example of this interaction is shown in Figure A-11. Core analyses provide data for identifying reservoir rock types, whereas well test studies assist in recognizing flow barriers and fractures. Simulation studies can be utilized to validate the geologic model against pressure-production performance. Often adjustments are required in the model to history match the actual performance.

Loudon Field Surfactant Pilot

Loudon field surfactant flooding pilot study details information required for simulation, and this work provides a comprehensive example to follow, as a case study. The pilot, located in central Illinois (see Figure B-2), contains five wells drilled on a 0.625-acre, unconfined five-spot pattern and an observation well midway between the producer and the northern most injector. Typical well logs and core description data for the pilot are shown in Figure B-3. The sandstone interval is the Weiler sandstone, which is the major reservoir unit at Loudon. The core-description graph records depositional environment and key rock units, oil staining, texture, lithology, sedimentary structures, and calcite cement of geologic origin. The logs and core-description data show that the

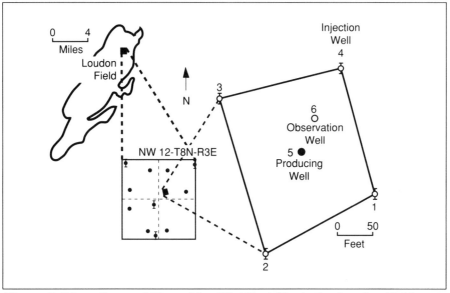

Figure B-2 Location and Plan of the Pilot-Test Site in the Loudon Field, Illinois (Copyright 1975, SPE, from JPT, May 1975[13])

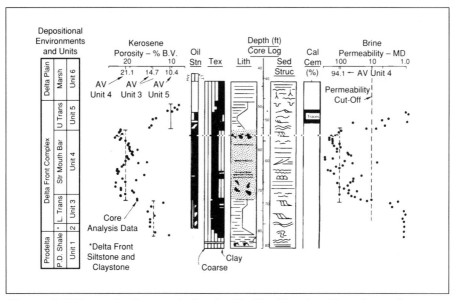

Figure B-3 Example Reservoir-Quality Profile Showing Porosity and Permeability Relationships to Lithology (Copyright 1975, SPE, from JPT, May 1975[13])

cored interval consists of six lithologic units. Representative photos of these intervals are shown in Figure B-4.

The environment of deposition provides the basis for rock framework studies, and it assists in understanding quality and three-dimensional distribution of the reservoir rock. The depositional conditions are recognized by examining slabbed cores (see Figure B-4). The subdivisions of the depositional framework in the pilot are indicated in Figure B-3. The three-dimensional relationships of the subdivisions illustrated in Figure B-5 indicate documented information from studies of modern deltas.

Core analysis data and well logs provide more definitive methods of reservoir characterization. It is helpful to obtain core analysis data on representative lithologies and prepare graphs of porosity vs. permeability as shown in Figure B-6. The rock type for each plug value is indicated to aid identification.

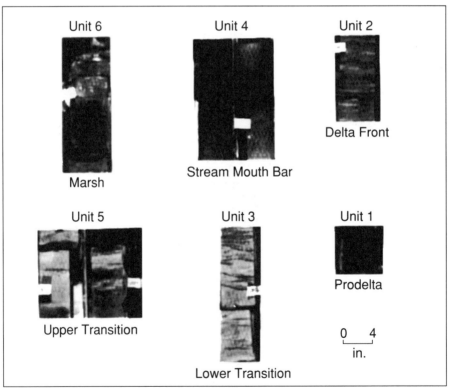

Figure B-4 Typical Photographs of the Six Lithologic Units (The photo for unit 3 is about 1 ft long) (Copyright 1975, SPE, from JPT, May 1975[13])

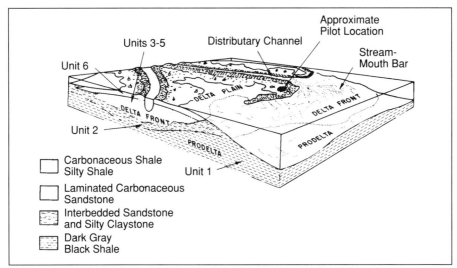

Figure B-5 Block Diagram Showing the Vertical and Areal Distribution of Units in a Typical Modern Delta (Copyright 1975, SPE, from JPT, May 1975[13])

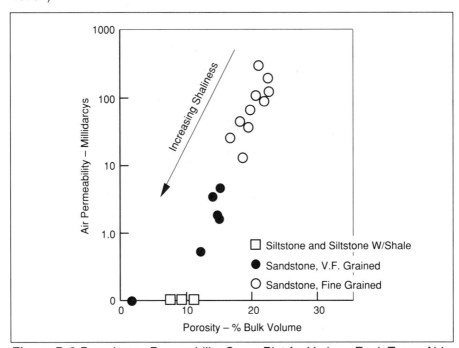

Figure B-6 Porosity vs. Permeability Cross-Plot for Various Rock Types Aids Identification of Reservoir Rocks (Copyright 1975, SPE, from JPT, May 1975[13])

Regional subsurface and surface studies demonstrated that the Loudon field is located on the flank of a large, ancient river system feeding a coastal deltaic complex (see Figure B-7). Most of the Loudon field, situated in the thinner, discontinuous delta front, was produced initially by solution gas drive. Later, waterflooding operations were begun to increase oil recovery.

To ascertain the reservoir continuing in a field, well logs and seismic section are analyzed. The objective of correlation studies is to determine the vertical and areal limits of reservoir. When cores are lacking, the depositional units often can be estimated by comparing well-log curves in cored and uncored sections.

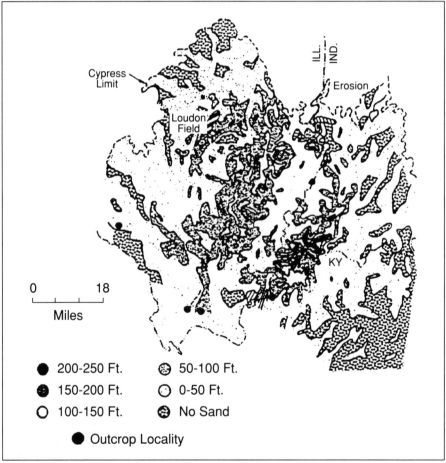

Figure B-7 Regional Thickness Map of the Chester (Weiler) Sandstone Reflects Depositional Controls on Sand Distribution (Copyright 1975, SPE, from JPT, May 1975[13])

The method used in the case study is shown in Figure B-8. Wells in the pilot and in the 16-acre area surrounding the pilot are correlated. The system used recognized points of depositional association in various well locations that formed a series of "loops" on a map of a 40-acre area Key curve points are numbered and, on cross-sections, lines were drawn to connect these points.

The pattern of lines in the interval above and below the Weiler, together with the depositional units (see Figure B-3), aided in determining the vertical and lateral limits of reservoir units in various wells.

Shale breaks of two scales were identified in the pilot (see Figures B-3 and B-9): (1) those with radii of at least 200 ft. but less than 1,000 ft. and (2) those with radii less than 50 ft. Tight streaks of carbonate cement occurred in five wells, but they were not correlative between wells.

In summary, framework studies are aimed at determining the number and the distribution of reservoir zones. This is accomplished

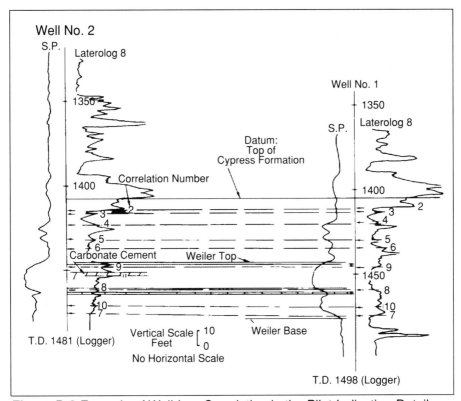

Figure B-8 Example of Well-Log Correlation in the Pilot Indicating Detail Required for Sand Continuity Studies (Copyright 1975, SPE, from JPT, May 1975[13])

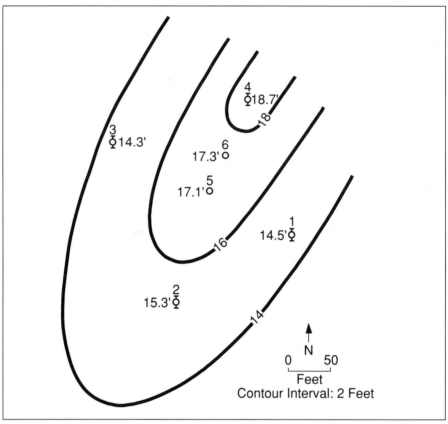

Figure B-9 Thickness Map of the Reservoir and Nonreservoir Rocks in Unit 4 (Copyright 1975, SPE, from JPT, May 1975[13])

with core-description graphs, well-log correlations, and structure and gross-thickness maps.

For each reservoir zone, it is necessary to establish the vertical and lateral distribution of pay, porosity, and permeability values. Laboratory analysis suggested that 10 md was the practical lower limit for effective surfactant flooding, and it was used as a cutoff value for determining net sand.

Net sand counts and preliminary evaluation of variations in porosity and permeability were developed, as shown in Figure B-3. Also shown in this figure is the permeability cutoff for determining pay.

If core plugs are widely spaced, then the rock interval between plugs can be estimated by spreading the plug values to include intervals with a similar lithology. Conventional log-analysis data, such as porosity or water saturation, can be used. The results from either of these methods can be compared with well test data. In the pilot, porosity was fairly constant, but permeability showed some variability.

Next, reservoir maps of pore volume and permeability capacity are developed. Such maps are important because they represent the three-dimensional attributes of certain parameters defining the physical model. Thus, these maps point up critical areas that may require careful evaluation during the construction of a simulation model. Pore volume and other maps can be constructed by cross-contouring a net-sand map with a porosity map (or a map of another parameter).

One of the most important steps in the integration study is verifying the geologic model against the actual performance, also known as "history matching." This step often requires adjusting some of the parameters of the reservoir model. However, any adjustments should be made cautiously and within reason by involving geoscientists and engineers. In the pilot case, a good match of tracer response was achieved. The simulated pore volume was 16,500 bbls; whereas the geologic model calculated value was 16,250 bbls.

Seismic Data

Three-dimensional seismic surveys help identify reserves that may not be produced optimally. The analysis can save costs by minimizing dry holes and poor producers. A 3-D survey shot during the evaluation phase is used to assist in the design of the development plan. With the development and production, data are constantly being evaluated to form the basis for locating production and injection wells, managing pressure maintenance, and performing workovers. These activities generate new information (logs, cores, DST's, etc.) that change maps, revise structure, and alter the reservoir stratigraphic model.

A 3-D seismic survey impacts the original development plan. With the drilling of development wells, the added information is used to refine the original interpretation. As time passes and the data builds, elements of the 3-D data that were initially ambiguous begin to make sense. The usefulness of a 3-D seismic survey lasts for the life of a reservoir.

A 3-D seismic data can be used to assist in (1) defining the geometric framework, (2) qualitative and quantitative definition of rock and fluid properties, and (3) flow surveillance.

Geostatistics

Haldorsen and Damsleth explain the use of reservoir-description techniques:

> A reservoir is intrinsically deterministic. It exists; it has potentially measurable, deterministic properties and features at all scales; and it is the end product of many complex processes ... that occurred over millions of years. Reservoir description is a combination of observations (the deterministic component), educated aiming (geology, sedimentology, [and the depositional environment]) and formalized "guessing" (the stochastic component).[20]

They further explain that stochastic techniques are applied to describe deterministic reservoirs because of (1) incomplete information about the reservoir on all scales, (2) complex spatial deposition of facies, (3) variability of rock properties, (4) unknown relationships between properties, (5) the relative abundance of singular pieces of information from wells, and (6) convenience and speed.

"Stochastic modeling" refers to the generation of synthetic geologic properties in one, two, or three dimensions. A number of plausible solutions can be created and simulated, and those results can be compared to see their effect on history matching.

Recent interest has focused on the use of fractal geostatistics and other stochastic techniques to map the variability of uncertainty associated with heterogeneous reservoirs.[21-23] Fractal geostatistics is based on the assumption that fractal statistics can be used to represent reservoir heterogeneity between wells as a random fractal variation superimposed on a smooth interpolation of correlated well log values.[11] Standard fractal statistics are used to determine the characteristics of the random fractal variation from analyses of well-log and core data. This approach is successful because the variation of properties of many natural systems is fractal in nature.

Geostatistics/stochastic modeling of reservoir heterogeneities is playing an important role in generating more accurate reservoir models. It provides a set of spatial data analysis tools as probabilistic language to be shared by geologists, geophysicists, and reservoir engineers, and it's a vehicle for integrating various sources of uncertain information.[23]

Geostatistics is useful in modeling the spatial variability of reservoir properties and the correlation between related properties such as porosity and seismic velocity. These models can then be used in the construction of numerical models for interpolating a property whose average is critically important and stochastic simulations for a property whose extremes are critically important. Geostatistics enable geologists to put their valuable information in a format that can be used by reservoir engineers. In the absence of sufficient data in a particular reservoir, statistical characteristics from other more mature fields in similar geologic environments and/or outcrop studies are utilized. By capturing the critical heterogeneities in a quantitative form, a more realistic geologic description is created.

Variograms (and correlograms) are used to study spatial continuity of a particular variable. Also, they can be applied to study cross-continuity of different variables at different locations. For example, one could compare porosity at a particular location to travel time at a nearby location. Once modeled, this spatial cross-correlation can be used in a multivariate regression procedure known as "cokriging" for building a porosity map not only from the available porosity data, but also from the seismic information.

Any unsampled value (e.g., porosity) can be estimated by generalized regression from surrounding measurements for the same value once the statistical relationship between the unknown being estimated and the available sample values has been defined. This prior model of the statistical similarity between data values is defined by correlogram. This generalized regression can also include nearby measurements of some different variable—seismic travel time or a facies code. When using such secondary information, one also needs a prior model of the cross-correlogram, which provides information on the statistical similarity between different variables at different locations. These generalized regression algorithms are collectively known as "kriging" and "cokriging".

Integrating widely different types of data, such as qualitative geological information and direct laboratory measurements, is not easy to achieve by multivariate regression or "cokriging". At a particular location where a porosity measurement does not yet exist, a consideration of the lithofacies may provide a reasonable range within which the unknown value should fall. In addition, if we have a porosity histogram developed for the lithofacies in question, we then can tell if the unknown porosity is more likely to be on the low end or the high end of the range.

The result of this updating is a probability distribution that provides the probability for the unsampled value to belong within any given class of values. For such a distribution, any optimal estimate for the unsampled value can be derived once an optimality criterion (e.g., least square error criterion) has been specified. Geostatistics is used in building probability distributions that characterize the present uncertainty about a reservoir's rock and fluid properties. If there are variable probability distributions representing the uncertainties in rock and fluid properties at different locations, each drawing from such a multivariate probability distribution represents a stochastic image of the reservoir. Such stochastic images honor all of the available information entered through the correlograms and cross-correlograms, yet they are different one from the other.

The differences between these stochastic images provide a direct usable visualization of the uncertainty about the reservoir rock and fluid properties. Where all the different outcomes agree, there is little or no uncertainty; where they differ most, there is maximum uncertainty.

Stochastic images represent alternative, equiprobable numerical models of the reservoir rock and fluid properties, and they can be used as input to flow simulators for sensitivity analysis.

To illustrate an application of the geostatistical concepts, porosity and permeability distributions for a reservoir were generated with fractal geostatistics based on the work of Hewett and Behrens.[24] For this field, the interwell permeabilities were determined with a porosity vs. permeability correlation, and the porosity distributions were obtained from the geostatistics.[25] Four different porosity/permeability representations were created by honoring any permeability barriers that existed. The resulting permeability distributions for this model are shown in Figure B-10. The result of this procedure was to generate four different geologic models of varying complexities for this reservoir.

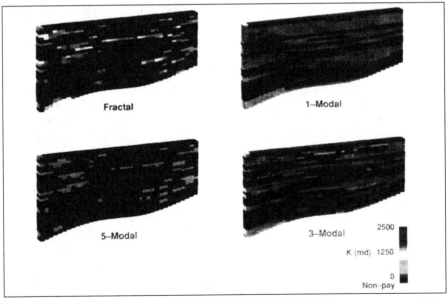

Figure B-10 Permeability Representations: N/T Model B (Copyright 1992, SPE, from SPERE, Sept. 1992[24] and copyright 1992, SPE, from JPT, Dec. 1992[25])

Engineering

Optimizing oil and gas recovery requires the following six steps:

1. Identify and define all individual reservoirs in a given field and their physical properties
2. Deduce past and predict reservoir performance
3. Minimize drilling of unnecessary wells
4. Define and modify (if necessary) wellbore and surface systems
5. Initiate operating controls at the proper time
6. Consider all pertinent economic and legal factors

After identifying the geologic model, additional engineering/production information are input in the model. These include reservoir fluid and rock properties, well location and completion, well-test pressures and pulse-test responses to determine well continuity and effective permeability. The use of material balance method permits the calculation of original hydrocarbon in place, aquifer influx, and areal extent of the reservoir. The use of injection/production profiles provide vertical fluid distribution.

Integration

Traditionally, data of different types have been processed separately, leading to several different models—a geologic model, a geophysical model, and a production/engineering model. Indicator geostatistics provide an approach to merge all of the relevant information and then produce reservoir models consistent with that information.

The importance of geology to the prediction of reservoir performance is recognized by reservoir engineers. However, it is important that the geologic (including the qualitative inferences) picture be transferred into a simulation model and not be time-consuming and frustrating.

Three-dimensional geological modeling programs have been developed to automate the generation of geological maps and cross-sections from exploration data.[12] Because these models are directly interfaced to the reservoir simulator, the reservoir engineer can easily utilize the complex reservoir description provided by the geologist for field development planning. In addition, the reservoir engineer can routinely and readily update their model with new data or interpretations and quickly provide consistent maps and sections for assessing results of activities like infill drilling. Three-dimensional geological modeling programs can provide maps and cross-sections in large numbers. This permits the engineer to become thoroughly familiar with the geology prior to designing the simulation model.

A revolution in simulation techniques has come with the advent of numerical simulation models. Today, the reservoir engineer seeks more data, both in quantity and detail, from the geologist and production engineer. On the other hand, the history matching of the reservoir can lead to a feedback of geological information to the geologist.

The degree of interaction between geoscientists and engineers has been well documented in the literature. Craig, et al. (1977)[26] and Harris and Hewitt (1977)[18] explained the value of synergism between engineering and geology. Craig, et al. emphasized the value of detailed reservoir description, utilizing geological, geophysical and reservoir simulation with the knowledge of geophysical tools, to provide a more accurate reservoir description for use in engineering calculations. Harris and Hewitt presented a geologic perspective of the synergism in reservoir management. They explained the reservoir inhomogeneity due to complex variations of reservoir continuity, thickness patterns, and pore-space properties (e.g., porosity permeability and capillary pressure).

A major breakthrough in reservoir modeling has occurred with the advent of integrated geoscience (reservoir description) and engineering (reservoir production performance) software designed to manage reservoirs more effectively and efficiently (see Figure B-11 and B-12). Several service, software and consulting companies are now developing and marketing integrated software installed in a common platform. These interactive and user-friendly software provides more realistic reservoir models. The users from different disciplines can work with the software cooperatively as a basketball team rather than passing their own results like batons in a relay race.

Case Studies

North Sea Lemen Field

As one early example, Craig, et al. used a multidisciplinary approach on the East Unit of the North Sea Lemen field from the time the field came on stream in 1968.[26] The field contained more than 10 Tcf gas—then the world's largest producing offshore gas field.

Working in a complex fault system, the company's reservoir engineers worked closely with geologists to "produce an accurate *a priori* reservoir description". The team tested the description against field per-

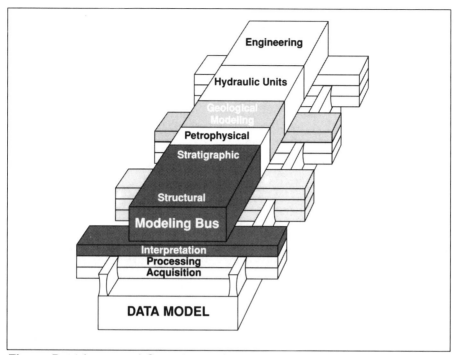

Figure B-11 Integrated Geoscience-Engineering Software System Concept (Courtesy of Western Atlas Software)

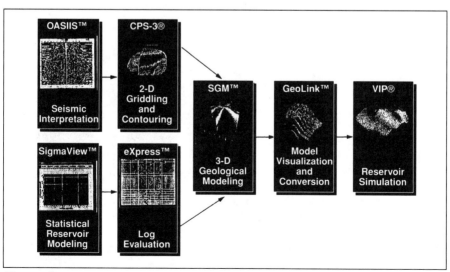

Figure B-12 Integrated Geoscience-Engineering Software System (Courtesy of Western Atlas Software)

formance in a 2D fine-grid, single-phase model and refined it with measured pressures from the first six years of production.

Geologists reviewed the location of faults and reservoir boundaries of the historical map. The resulting model successfully predicted pressure for an additional two years. The proven accuracy of the model led to confident planning of future platform and compression requirements, providing more than three years' lead time to install equipment.

Brassey Oil Fields

Recently Woofer and MacGillivary[27] presented a case study of the development of an aeolian sand, Brassey oil field, British Columbia. The case study presented an aggressive engineering/geoscience team approach to prepare development plans. Production and miscible injection commenced simultaneously in 1989, only two years after discovery. The integration of the reservoir description, volumetrics, seismic delineation, and well-test data provided the basis for determining reservoir continuity and size.

A geologic model was developed before the miscible flood was designed and concurrent with seismic delineation. Excellent core control throughout the field aided detailed sedimentologic and petrographic interpretations, but the wide well spacing of 0.5 mile or more limited interwell correlations. A geologic model was developed with the aid of geophysical and engineering information.

Reservoir simulations were constructed with the geologic model. Although the well spacing and thinness of the reservoir limited interwell correlations to two layers, additional model sensitivities were run with more layers on the basis of case permeability variations.

Teamwork was critical to the development of the reservoir model. The geophysicist defined the pool edges; the engineers described pools by using well test pressures and pulse-test responses to determine well continuity; and the geologist mapped the reservoir properties. The team's goal was to continue to improve the model as production history and tracer information became available.

Original oil in place was calculated from material-balance analysis and derived from a hand-contoured volume. Material-balance calculations show 22 to 27 MMSTB, compared with 24 MMSTB of oil from mapped volumes. Agreement of independently calculated volumes gave the confidence to proceed with field development, and it was the basis for the predicted field recovery and producing allowable.

References

1. Satter, A., J. E. Varnon, and M. T. Huang. "Reservoir Management Technical Perspective." SPE Paper 22350, SPE International Meeting on Petroleum Engineering, Beijing, China, March 24-27, 1992.
2. Thomas, G. W. "The Role of Reservoir Simulation in Optimal Reservoir Management." SPE Paper 14129, International Meeting on Petroleum Engineering, Beijing, China, March 17-20, 1986.
3. Saleri, N. G. and R. M. Toronyi. "Engineering Control in Reservoir Simulation." SPE Paper 18305, SPE 63rd Annual Technical Conf. & Exhib., Houston, TX, October 2-5, 1988.
4. Thakur, G. C. "What is Reservoir Management?" *JPT* (June 1996): 520-525.
5. Satter, A., D. F. Frizzell and J. E. Varnon. "The Role of Mini-Simulation in Reservoir Management." SPE Paper 22370, International Meeting on Petroleum Engineering, Beijing, China, March 24-27, 1992.
6. Coats, K. H. "Reservoir Simulation: State of the Art," *JPT* (August 1982): 1633-1642.
7. Coats, K. H. "Use and Misuse of Reservoir Simulation Models." SPE Reprint Series no. 11, Numerical Simulation, (1973): 183-190.
8. Breitenbach, E. A. "Reservoir Simulation: State of the Art," *JPT* (September 1991): 1033-1036.
9. Tang, R. W. et al. "Reservoir Studies With Geostatistics To Forecast Performance," SPE Reservoir Engineering (May 1991): 253-258.
10. Harris, D. E. and R. L. Perkins "A Case Study of Scaling Up 2D Geostatistical Models to a 3D Simulation Model." SPE Paper 22760 presented at the 1992 Annual Technical Conference and Exhibition, Dallas, TX, October 6-9, 1991.
11. Emanuel, A. S. et al. "Reservoir Performance Prediction Methods Based on Fractal Geostatistics," *SPERE* (August 1989): 311-318.
12. Johnson, C. R and T. A. Jones "Putting Geology Into Reservoir Simulations: A Three-Dimensional Modeling Approach." SPE Paper 18321 presented at the 1988 Annual Technical Conference and Exhibition, Houston, TX, October 2-5, 1988.
13. Harris, D. G. "The Role of Geology in Reservoir Simulation Studies," *JPT* (May 1975): 625-632.
14. Ghauri, W. K. et al "Changing Concepts in Carbonate Waterflooding—West Texas Denver Unit Project—An Illustrative Example," *JPT* (June 1974): 595-606.
15. Staggs, H. M. and E. F. Herbeck. "Reservoir Simulation Models—An Engineering Overview," JPT (December 1971): 1428-1436.
16. Robertson, J. D. "Reservoir Management "Using 3-D Seismic Data," Geophysics: The Leading Edge of Exploration (February 1989): 25-31.
17. Nolen-Hoeksema, R. C. "The Future of Geophysics in Reservoir Engineering," *Geophysics: The Leading Edge of Exploration* (December 1990): 89-97.

18. Harris, D. G. and C. H. Hewitt "Synergism in Reservoir Management—The Geologic Perspective," *JPT* (July 1977): 761-770.
19. Harris, D. G. "The Role of Geology in Reservoir Simulation Studies," *JPT* (May 1975): 625-632.
20. Haldorsen, H. H. and E. Damsleth. "Stochastic Modeling," *JPT* (April 1990): 404-412.
21. Hewitt, T. A. "Fractal Distributions of Reservoir Heterogeneity and Their Influence on Fluid Transport" SPE Paper 15386 presented at the SPE Annual Technical Conference and Exhibition, New Orleans, October 5-8, 1986.
22. Begg, S. H., R. R. Carter and P. Dranfield. "Assigning Effective Values to Simulator Gridblock Parameters for Heterogeneous Reservoirs," *SPERE* (November 1989): 455-463.
23. Journel, A. G. and F. G. Alabert. "New Method for Reservoir Mapping," *JPT* (February 1990): 212-218.
24. Hewett, T. A. and Behrens, R. A.: "Conditional Simulation of Reservoir Heterogeneity With Fractals," *SPERE* (September 1992): 217-225.
25. Saleri, N. G. et al. "Data and Data Hierarchy," *JPT* (December 1992): 1286-1293.
26. Craig, F. F. et al. "Optimized Recovery Through Continuing Interdisciplinary Cooperation," *JPT* (July 1977): 755-760.
27. Woofer, D. M. and J. MacGillivary. "Brassey Oil Field, British Columbia: Development of an Aeolian Sand—A Team Approach," *SPERE* (May 1992): 165-172.

APPENDIX C

Data Acquisition, Analysis and Management

An efficient data management program consisting of acquisition, analysis, validating, storing and retrieving plays a key role in reservoir management. It requires planning, justification, prioritizing and timing.

Chapter 4 presents specific data required for waterflood asset management. This Appendix presents a general discussion of data types, data acquisition and analysis, data validation, data storing and retrieval, data applications, and examples.

Data Types[1,2,3]

Throughout the life of a reservoir, from exploration to abandonment, an enormous amount of multidisciplinary data are collected (see Table C-1). Also included in this table is the professionals responsible for acquisition and analyses. It is emphasized that the multidisciplinary professionals need to work as an integrated team to develop and implement an efficient data management program.

Data Acquisition and Analysis[2,3,4]

Multi-disciplinary groups, i.e., geophysicists, geologists, petrophysicists, drilling, reservoir, production and facilities engineers are involved in collecting various types of data throughout the life of the reservoirs. Land and legal professionals also contribute to the data collection

Table C-1 Reservoir Data

Life Cycle/Type	Data	Responsibility
Exploration		
Seismic	Structure, stratigraphy, faults, bed thickness, fluids, interwell heterogeneity	Land men, geophysicists, seismologists
Geological	Depositional environment diagenesis, lithology, structure, faults, and fractures	Geologists
Discovery		
Drilling	Mud, cuttings, geolograph	Drilling engineer, geologist
Logging	Depth, lithology, thickness, porosity, fluid saturation, gas/oil, water/oil and gas/water contacts	Development geologists, petrophysicists, and drilling engineers
Fluids (oil, water, and gas)	Formation volume factors, compressibilities, viscosities, solubilities, chemical compositions, phase behavior, and specific gravities	Reservoir engineers and laboratory analysts
Well Test	Reservoir pressure, effective permeability, thickness, presence of fractures or faults, productivity index	Reservoir and production engineers
Delineation		
Coring		Geologists, drilling and reservoir engineers, laboratory analysts
Basic	Depth, lithology, thickness, porosity, permeability, and residual fluid saturation	
Special	Relative permeability, capillary pressure, pore compressibility, grain size, and pore size distribution	
Logging	Same as above	Development geologists, petrophysicists, and drilling engineers
Primary Production		
Production	Oil, water and gas productions	Production engineers
Fluids as desired	Same as above	Reservoir engineers, laboratory analysts
Well Test as desired	As above, reservoir continuity	Reservoir, production engineers
Waterflooding		
Production	Same as above, production profiles	Production engineer
Injection	Water injection, water quality	Production engineer
Well test	As above as desired, injection and production profiles	Reservoir and production engineers

process. Most of the data, except for the production and injection data are collected during delineation and development of the fields.

An effective data acquisition and analysis program requires careful planning and well coordinated team efforts of multi-disciplinary geoscientists and engineers throughout the life of the reservoir. On one hand, there may be the temptation to collect lots of data, and on the other hand, to short-cut data acquisition to reduce costs.[4] Justification, priority, timeliness, quality, and cost-effectiveness should be the guiding factors in data acquisition and analysis. It will be more effective to justify to management data collection, if the need for the data, costs and benefits are clearly defined.

Dandona[3] reminds that certain types of data such as core derived information, initial fluid properties, fluid contacts, and initial reservoir pressures can only be obtained at an early development stage. Coring, logging, and initial reservoir fluid sampling should be made at appropriate times using proper procedure and analyses. Normally, all wells are logged; however, an adequate number of wells should be cored to validate the log data. Initial bottom hole pressure measurements should be made preferably at each well and at selected "key wells" periodically. According to Woods and Abib[2], key wells represent 25% of the total wells. Also, they found it beneficial to measure pressures in all wells at least every two to three years to aid in calibrating reservoir models.

3-D seismic data can be collected from mature fields during production for better reservoir definition. Cross-well tomography can provide well connectivity and better reservoir characterization.

Laboratory rock properties, such as oil-water and gas-oil relative permeabilities, and fluid properties, such as PVT data, are not always available. Empirical correlations can be used to generate these data.[5-15]

It is essential to establish specification of what and how much data to be gathered, and the procedure and frequency to be followed. A logical, methodical, and sequential data acquisition and analysis program suggested by Raza is shown in Table C-2.[4]

Data Validation[2,4]

Field data are subjected to many errors, i.e., sampling, systematic, random, etc. Therefore, the collected data needs to be carefully reviewed and checked for accuracy as well as for consistency.

In order to assess validity, core and log analyses data should be carefully correlated and their frequency distributions made for identifying different geologic facies. Log data should be carefully calibrated using core data for porosity and saturation distributions, net sand deter-

Table C-2 An Efficient Data Flow Diagram (Copyright 1992, SPE, from JPT, April 1992)

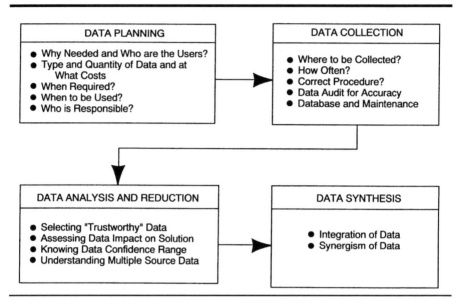

mination and geological zonation of the reservoir. The reservoir fluid properties can be validated by using equation of state calculations and by empirical correlations.

The reasonableness of geological maps should be established by using the knowledge of depositional environment. The presence of faults and flow discontinuities as evidenced in a geological study can be investigated and validated by pressure interference and pulse and tracer tests.

The reservoir performance should be closely monitored while collecting routine production and injection data including reservoir pressures. If past production and pressure data are available, classical material balance techniques and reservoir modeling can be very useful to validate the volumetric original hydrocarbons-in-place, and aquifer size and strength.

Data Storing and Retrieval

The reconciled and validated data from the various sources need to be stored in a common computer database accessible to all interdisciplinary end users. As new geoscience and engineering data are available, the database requires updating. The stored data are used to carry out multipurpose reservoir management functions including monitoring and evaluating the reservoir performance.

Storing and retrieval of data during reservoir life cycle poses a major challenge in the petroleum industry today. The problems are:

- Incompatibility of the software and data sets from the different disciplines
- Databases usually do not communicate with each other

Many oil companies are staging an integrated approach to solving these problems.[16] In late 1990, several major domestic and foreign oil companies formed Petrochemical Open Software Corporation (POSC) to establish industry standards and a common set of rules for applications and data systems within the industry. POSC's technical objective is to provide a common set of specifications for computing systems which will allow data to flow smoothly between products from different organizations and will allow users to move smoothly from one application to another.

Data Application

A better representation of the reservoir is made from 3-D seismic information. The cross-well tomography provides interwell heterogeneity.

Geological maps such as gross and net pay thickness, porosity, permeability, saturation, structure, and cross-section are prepared from seismic, core and log analysis data. These maps which also include faults, oil-water, gas-water and gas-oil contacts are used for reservoir delineation, reservoir characterization, well locations and estimates of original oil and gas in place.

The more commonly used logging systems are:

- *Open Hole Logs:*

    Resistivity, Induction, Spontaneous Potential, Gamma Ray, Density, Sonic Compensated Neutron, Sidewall Neutron Porosity, Dielectric, and Caliper

- *Cased Hole Logs:*

    Gamma Ray, Neutron (except SNP), Carbon/Oxygen, Chlorine, Pulsed Neutron and Caliper

The well log data which provide the basic information needed for reservoir characterization, are used for mapping, perforations, estimates of original oil and gas in place, and evaluation of reservoir performance. Production logs can be used to identify remaining oil saturation in undeveloped zones in existing production and injection wells. Time-lapse logs in observation wells can detect saturation changes and fluid contact movement. Also, log-inject-log can be useful for measuring residual oil saturation.

Core analysis is classified into conventional, whole-core, and sidewall analyses. The most commonly used conventional or plug analysis involves the use of a plug or a relatively small sample of the core to represent an interval of the formation to be tested. Whole core analysis involves the use of most of the core containing fractures, vugs or erratic porosity development. Sidewall core analysis employs cores recovered by sidewall coring techniques.

Unlike log analysis, core analysis gives direct measurement of the formation properties, and the core data are used for calibrating well log data. These data can have a major impact on the estimates of hydrocarbon-in-place, production rates and ultimate recovery.

The fluid properties are determined in the laboratories using equilibrium flash or differential liberation tests. The fluid samples can be either subsurface sample or a recombination of surface samples from separators and stock tanks. Fluid properties can be also estimates by using correlations.[10-15]

Fluid data are used for volumetric estimates of reservoir oil and gas, reservoir type, i.e., oil, gas or gas condensate, and reservoir performance analysis. Fluid properties are also needed for estimating reservoir performance, wellbore hydraulics, and flow line pressure losses.

The well test data are very useful for reservoir characterization and reservoir performance evaluation. Pressure build-up or falloff tests provide the best estimate of the effective permeability-thickness of the reservoir in addition to reservoir pressure, stratification, presence of faults and fractures. Pressure interference and pulse tests provide reservoir continuity and barrier information. Multi-well tracer tests used in waterflood and in enhanced oil recovery projects give the preferred flow paths between the injectors and producers. Single well tracer tests are used to determine residual oil saturation in waterflood reservoirs. Repeat formation tests can measure pressures in stratified reservoirs indicating varying degree of depletion in the various zones.

Production and injection data are needed for reservoir performance evaluation.

Table C-3 lists required data for the techniques used for analyzing reservoir performance, i.e., volumetric, decline curve, material balance, and simulation methods.

Table C-3 Data Required for Reservoir Performance Analyses (Satter and Thakur, G.C. *Integrated Petroleum Reservoir Management: A Team Approach*, Copyright PennWell Books, 1994)

Data Group	Volumetric	Decline Curve	Material Balance	Mathematical Models
Geometry	Area, thickness	No	Area, thickness Homogeneous	Area, thickness Heterogeneous
Rock	Porosity, saturation	No	Porosity, saturation, relative permeability, compressibility Homogeneous	Porosity, saturation, relative permeability, compressibility, capillary pressure Heterogeneous
Fluid	Form. vol. factors	No	PVT Homogeneous	PVT Heterogeneous
Well	No	No	PI for rate vs. time	Locations Perforations PI
Production & Injection	No	Production	Yes	Yes
Pressure	No	No	Yes	Yes

Example Data

Engineering studies of reservoirs G-1, G-2, and G-3 of Meren Field offshore Nigeria were made by Thakur, et al.[17] The studies included geologic evaluation, material balance calculations, and 3-phase, 3-D reservoir performance simulations. Geological, reservoir and production data used for the studies are presented in Figures C-1 to C-8 and Tables C-4 and C-5.

Additional rock and fluid properties of Meren G-1/G-2 reservoirs can be calculated using correlations. The calculated data are presented in Figures C-9 to C-11.

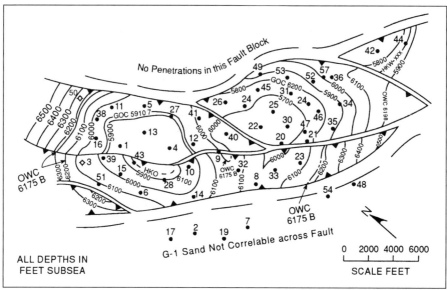

Figure C-1 Structure Map, Sand G-1 (Copyright 1982, SPE, from JPT, April 1982)

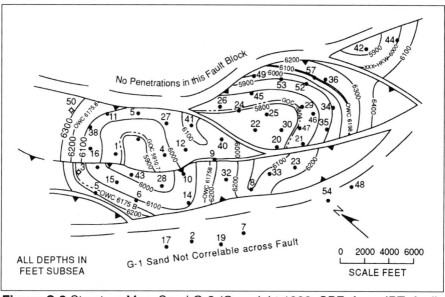

Figure C-2 Structure Map, Sand G-2 (Copyright 1982, SPE, from JPT, April 1982)

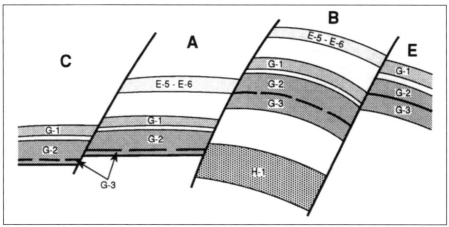

Figure C-3 Meren Field Cross Section Showing Sand G Juxtaposition (Copyright 1982, SPE, from JPT, April 1982)

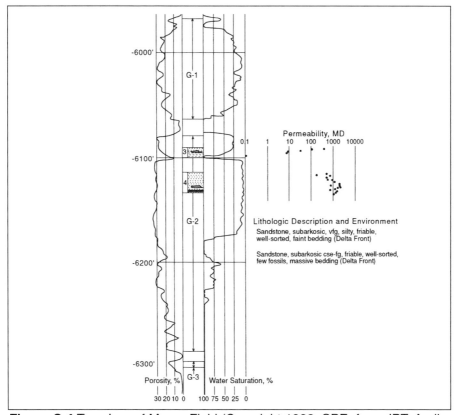

Figure C-4 Type Log of Meren Field (Copyright 1982, SPE, from JPT, April 1982)

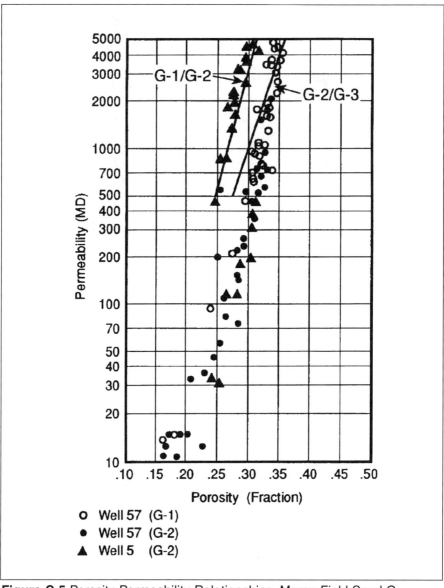

Figure C-5 Porosity-Permeability Relationships, Meren Field Sand G
(Copyright 1982, SPE, from JPT, April 1982)

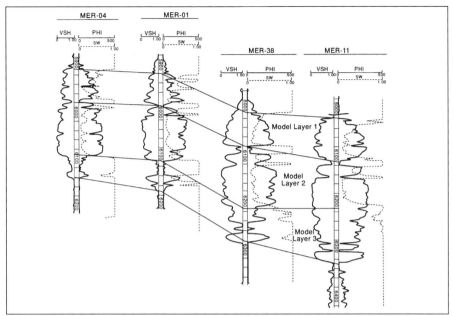

Figure C-6 Cross Section A-A' (Copyright 1982, SPE, from JPT, April 1982)

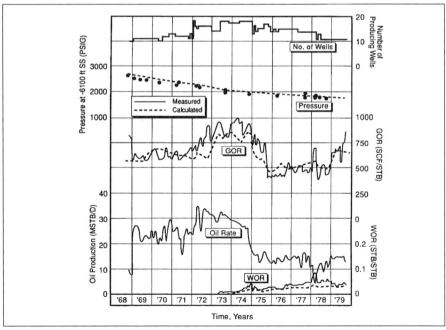

Figure C-7 Reservoir History Match, Sands G-1/G-2 (Copyright 1982, SPE, from JPT, April 1982)

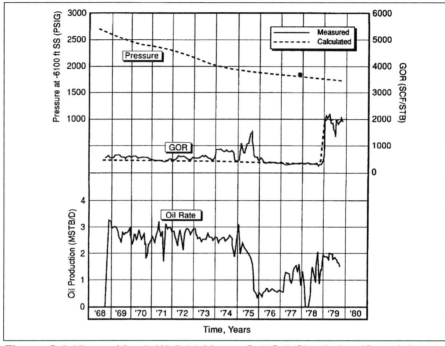

Figure C-8 History Match-Well 11 Meren G-1/G-2 Simulation (Copyright 1982, SPE, from JPT, April 1982)

Table C-4 Basic Reservoir Data

	Reservoir	
	G-1/G-2	**G-2/G-3**
Datum depth, ft subsea	-6,100	-6,000
Rock type	sandstone	sandstone
Average thickness, ft	138	126
Average porosity, %	27	32
Average permeability, md	1,150	1,775
Average connate water saturation, %	24	14
Initial reservoir pressure (at datum), psig	2,660	2,560
Average bubble-point pressure, psig	2,629	2,500
Initial oil volume factor, RB/STB	1,327	1,312
Initial solution GOR, scf/STB	566	588
Initial oil viscosity, cp	0.575	0.460
Oil gravity, °API	34	33
Initial oil/water contact, ft subsea	-6,175	-6,197
Initial gas/oil contact, ft subsea	-6,000	-5,804
OOIP, MMSTB	281.5	276.8
Original gas in place, Bscf	205.6	176.7

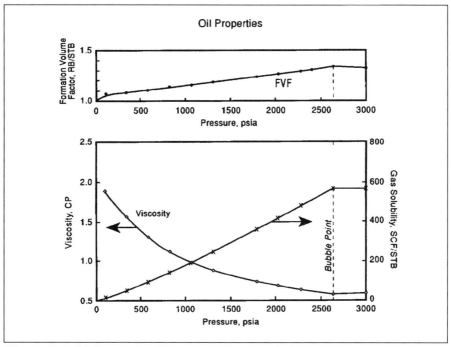

Figure C-9 Oil Properties (Satter and Thakur, G.C. *Integrated Petroleum Reservoir Management: A Team Approach*, Copyright PennWell Books, 1994)

Table C-5 Reservoir Fluid Properties Pressure Range of Saturation Pressure to 1,500 psig

	Reservoir	
	G-1/G-2	G-2/G-3
Oil volume factor, RB/STB	0.327 to 1.245	1.312 to 1.214
Reciprocal gas volume factor, scf/RB	910 to 475	833 to 490
Water volume factor, RB/STB	1.048 to 1.056	1.049 to 1.056
Solution GOR, scf/STB	566 to 368	588 to 360
Oil phase viscosity, cp	0.575 to 0.751	0.460 to 0.633
Oil phase pressure gradient, psi/ft	0.3068 to 0.3211	0.3201 to 0.3340
Gas viscosity, cp	0.0184 to 0.0146	0.0195 to 0.0142
Gas gradient, psi/ft	0.0616 to 0.0319	0.0536 to 0.0323

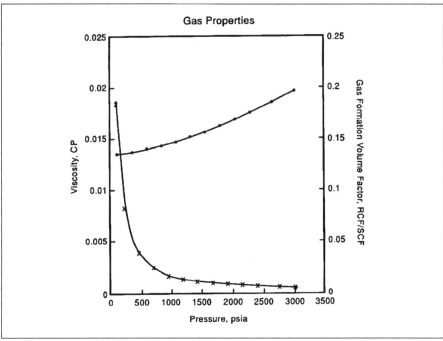

Figure C-10 Gas Properties (Satter and Thakur, G.C. *Integrated Petroleum Reservoir Management: A Team Approach*, Copyright PennWell Books, 1994)

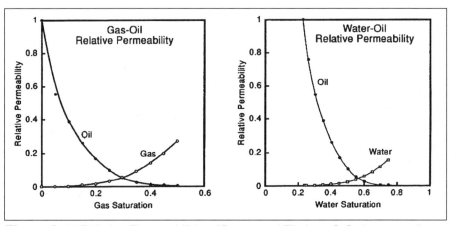

Figure C-11 Relative Permeabilities (Satter and Thakur, G.C. *Integrated Petroleum Reservoir Management: A Team Approach*, Copyright PennWell Books, 1994)

References

1. Satter, A., Varnon, J. E., and Hoang, M. T.: "Integrated Reservoir Management," *JPT* (Dec.1994).
2. Woods, E. G. and Abib, Osmar: "Integrated Reservoir Management Concepts," Reservoir Management Practices Seminar, SPE Gulf Coast Section, Houston, Texas, May 29, 1992.
3. Dandona, A. K., Alston, R. B., and Braun, R. W.: "Defining Data Requirements for a Simulation Study," SPE Paper 22357, SPE International Meeting on Petroleum Engineering, Beijing, China, March 24-27, 1992.
4. Raza, S. H.: "Data Acquisition and Analysis: Foundational to Efficient Reservoir Management", *JPT* (April, 1992).
5. Corey, A. T.: *Prod. Monthly* (1954) 19, No. 1, 38.
6. Wyllie, M. R. J. and Gardner, G. H. F.: "Generalized Kozeny-Carmen Equation, Parts 1 and 2," *World Oil* (1958) 146, No. 4 and No. 5.
7. Stone, H. L.: "Estimation of Three-Phase Oil Relative Permeability," *JPT* (1970), 214-218.
8. Dietrich, J. K. and Bondor, P. L.: "Three-Phase Oil Relative Permeability Models," SPE 6044 paper presented in the SPE Annual Meeting in New Orleans, October, 1976.
9. Honarpour, M., Koederitz, L. F. and Harvey, H. A.: "Empirical Equations for Estimating Two-Phase Relative Permeability in Consolidated Rock," *JPT* (December 1982), 2905-2908.
10. Beal, C.: "The Viscosity of Air, Water, Natural Gas, Crude Oil and its Associated Gases at Oil Field Temperature and Pressure," *Trans. AIME* (1946) 94.
11. Standing M. B.: "A Generalized Pressure-Volume-Temperature Correlation for Mixture of California Oils and Gases," Drilling and Production Practice API (1947) 275.
12. Lasater, J. A.: "Bubble Point Pressure Correlation," *Trans. AIME* (1958) 379.
13. Chew, J. and Connally, C. A.: "A Viscosity Correlation for Gas Saturated Crude Oils," *Trans. AIME* (1959) 20.
14. McCain, W. D.: "The Properties of Petroleum Fluids," Petroleum Publishing Co,. Tulsa, Oklahoma (1973).
15. Beggs, H. D. and Robinson, J. R.: "Estimating the Viscosity of Crude Oil Systems," *JPT* (Sept. 1975) 1140.
16. Johnson, J. P.: "POSC Seeking Industry Software Standards, Smooth Data Exchange," *Oil & Gas J.* (October 26, 1992).
17. Thakur, G. C., et al.: "Reservoir Studies of G-1, G-2, and G-3 Reservoirs, Meren Field, Nigeria," *JPT* (April 1982), 721-732.

APPENDIX
D

Reservoir Engineering Aspects Of Waterflooding

This appendix presents a review of reservoir engineering aspects of waterflooding[1,2] as follows:

- Immiscible Displacement Theory
- Flood Pattern
- Reservoir Heterogeneity
- Recovery Efficiency
- Injection Rates

Immiscible Displacement Theory

Darcy's law as shown below is the basic equation to describe the flow of fluids through porous media:

$$q = - \frac{kA}{\mu} \left(\frac{dp}{ds} - \frac{\rho g}{1.0133} \frac{dz}{ds} \; x \; 10^{-6} \right) \qquad \text{(D-1)}$$

where:

A = the cross-sectional area of rock and pore, in the direction of flow, cm^2

dp / ds = pressure gradient along the direction of flow, atm/cm

dz / ds = gradient in the vertical direction

g = acceleration due to gravity, cm/sec^2 (= 980.7 cm/sec^2)

k = permeability, darcies

μ = viscosity of flowing fluid, centipoise (cp)
ρ = density of the flowing fluid, g/cm^3
q = flow of fluid, cm^3/sec

Displacement of oil from a porous medium by immiscible water can be described by the fractional flow equation, and the frontal advance theory.[3,4]

Applying Darcy's law separately to oil and water flows, and considering viscous, gravitational, and capillary effects, the fractional flow equation of water displacing oil in practical units is:

$$f_w = \frac{1 + 0.001127 \dfrac{kk_{ro}}{\mu_o} \dfrac{A}{q_t} \left[\dfrac{\partial p_c}{\partial L} - \Delta \rho \sin \alpha_d \right]}{1 + \dfrac{\mu_w}{\mu_o} \dfrac{k_{ro}}{k_{rw}}} \qquad (D-2)$$

where:

A = area, sq. ft.
f_w = fraction of water flowing
k = absolute permeability, millidarcy
k_{ro} = relative permeability to oil
k_{rw} = relative permeability to water
μ_o = oil viscosity, cp
μ_w = water viscosity, cp
L = distance along direction of flow, ft
p_c = capillary pressure = $p_o - p_{w'}$ psi
q_t = total flow rate = $q_o + q_{w'}$ B/day
$\Delta \rho$ = water-oil density difference = $p_w - p_{o'}$, gm/cc
α_d = angle of formation dip to the horizon, degree

The fractional flow of water for given rock and fluid properties, and flooding conditions is a function of water saturation only because the relative permeability and capillary pressure are functions of saturation only.

Neglecting gravity and capillary effects, the above fractional flow equation is reduced to:

$$f_w = \frac{1}{1 + \dfrac{\mu_w}{\mu_o} \dfrac{k_{ro}}{k_{rw}}} \qquad (D-3)$$

Using the oil water relative permeability data shown in Figure D-1 and an oil-water viscosity ratio of 2, calculated fractional flow curve is shown in Figure D-2.

The linear frontal advance equation for water, based upon conservation of mass and assuming incompressible fluids, is given by:

$$\left(\frac{\partial x}{\partial t}\right)_{sw} = \frac{q_t}{A\phi}\left(\frac{\partial f_w}{\partial S_w}\right)_t \tag{D-4}$$

The frontal advance equation can be used to derive the expressions for average water saturation as follows:

At breakthrough: $\quad \bar{S}_{wbt} - S_{wc} = \left(\frac{\partial S_w}{\partial f_w}\right)_f = \frac{S_{wf} - S_{wc}}{f_{wf}}$ $\tag{D-5}$

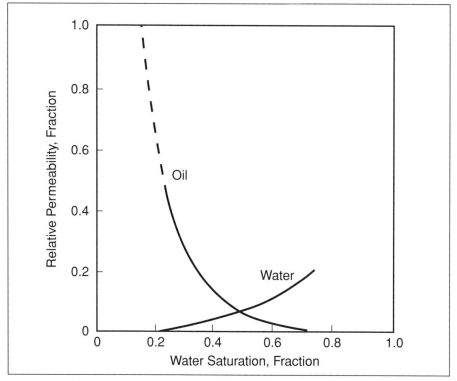

Figure D-1 Oil-Water Relative Permeabilities (Abdus Satter and Ganesh Thakur, Integrated Petroleum Reservoir Management: A Team Approach, Copyright PennWell Books, 1994)

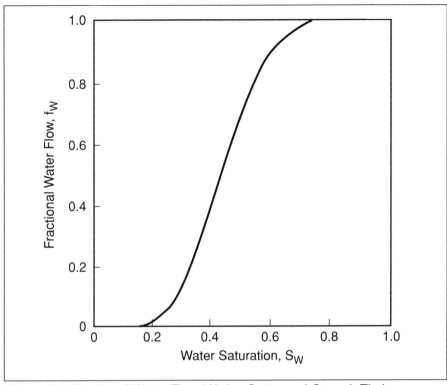

Figure D-2 Fractional Water Flow (Abdus Satter and Ganesh Thakur, *Integrated Petroleum Reservoir Management: A Team Approach*, Copyright PennWell Books, 1994)

After breakthrough: $\quad \overline{S}_w - S_{w2} = \dfrac{1 - f_{w2}}{\left(\dfrac{\alpha f_w}{\alpha S_w}\right) S_{w2}}$ $\qquad$ (D-6)

where:

f_{wf}	=	fraction of water flowing at the flood front
f_{w2}	=	fraction of water flowing at the producing end of the system
$\overline{S}_w$	=	average water saturation after breakthrough, fraction
S_{wf}	=	water saturation at the flood front, fraction
S_{wbt}	=	average water saturation at breakthrough, fraction
S_{wc}	=	connate water saturation, fraction
S_{w2}	=	water saturation at the producing end of the system, fraction

Figure D-3 presents graphical solutions for average water saturations at and after water breakthrough. The average water saturations can be used to calculate displacement efficiencies before and after water breakthrough.

Displacement efficiency that is governed by rock and fluid properties is given by:

$$E_D = \frac{\dfrac{S_{oi}}{B_{oi}} - \dfrac{S_{or}}{B_{or}}}{\dfrac{S_{oi}}{B_{oi}}} \qquad (D\text{-}7)$$

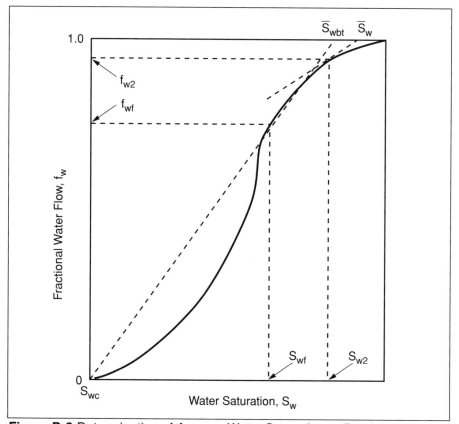

Figure D-3 Determination of Average Water Saturation at Breakthrough

where:

S_o = oil saturation, fraction
B_o = oil formation volume factor, RBO/STBO
i, r = subscripts denoting initial (before flooding) and residual (after flooding) condition, respectively

If oil and water are the only fluids present in the formation:

$$S_o = 1 2 S_w$$

Then assuming B_{or} is approximately equal to B_{oi}, oil displacement efficiency can be re-expressed as:

$$E_D = \frac{S_{wor} - S_{wi}}{1 - S_{wi}} \qquad \text{(D-8)}$$

when:

B_{oi} = B_{or}
S_{wor} = water saturation at the residual oil saturation which can be determined from the fractional flow curve for a given fractional water flow

Flood Pattern

The commonly used flood patterns, i.e., injection-production well arrangements are shown in Figure D-4, and their characteristics are given in Table D-1. Injectors positioned around the periphery of a reservoir, peripheral injection, and along the crest of small reservoirs with sharp structural features, crestal injection, are also used.

Reservoir Heterogeneity

Reservoirs are not uniform in their properties such as permeability, porosity, pore size distribution, wettability, connate water saturation, and fluid properties. The variations can be areal and vertical. The heterogeneity of the reservoirs is attributed to the depositional environments and subsequent events, as well as to the nature of the particles constituting the sediments. The performances of the reservoirs, whether primary or waterflood, are greatly influenced by their heterogeneities.

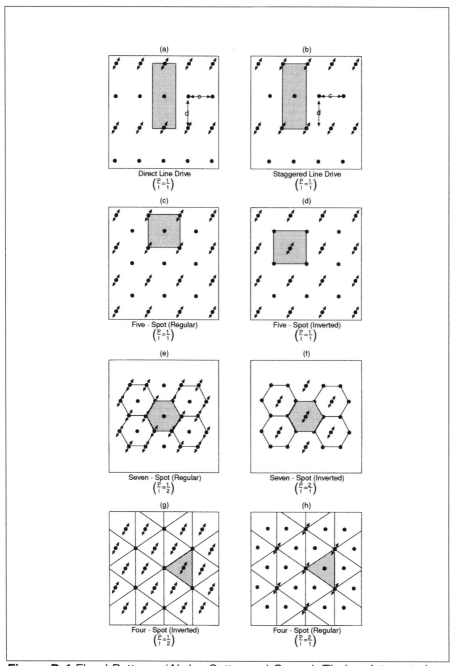

Figure D-4 Flood Patterns (Abdus Satter and Ganesh Thakur, Integrated Petroleum Reservoir Management: A Team Approach, Copyright PennWell Books, 1994)

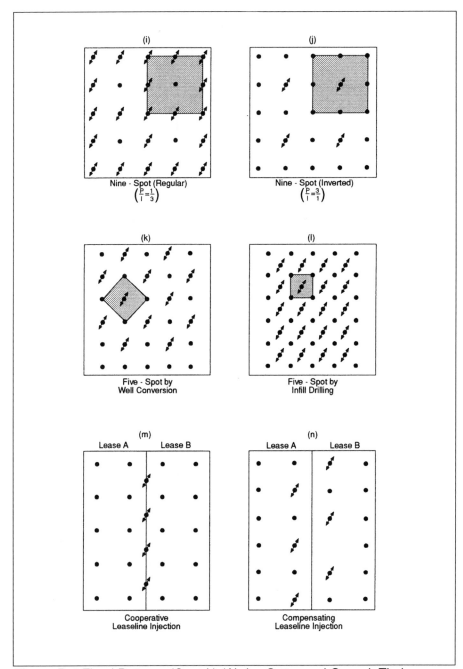

Figure D-4 Flood Patterns (Contd.) (Abdus Satter and Ganesh Thakur, Integrated Petroleum Reservoir Management: A Team Approach, Copyright PennWell Books, 1994)

Table D-1 Characteristics of Waterflood Patterns

Pattern	P/I Regular	P/I Inverted	d/a	EA, %
Direct Line Drive	1	–	1	56
Staggered Line Drive	1	–	1	78
4-Spot	2	1/2	0.866	–
5-Spot	1	1	1/2	72
7-Spot	1/2	2	0.866	–
9-Spot	1/3	3	1/2	~80

P	=	number of production wells
I	=	number of injection wells
d	=	distance from an injector to the line connecting two producing wells
a	=	distance between wells in line in regular pattern
EA	=	areal sweep efficiency at water breakthrough at a producing well for a water-oil mobility ratio = 1

Commonly used methods to characterize vertical permeability stratification are:

1. Flow capacity distribution (permeability x thickness), which is evaluated from a plot of the cumulative capacity versus cumulative thickness, of a reservoir having layered permeability (Figure 3-15). For a uniform permeability, the capacity distribution would plot as the straight line. Deviation from this straight line is a measure of the heterogeneity due to permeability variation

2. Lorenz coefficient which is based upon the flow capacity distribution is a measure of the contrast in permeability from the homogeneous case. It is defined by the ratio of the area ABCA to the area ADCA (Figure 3-15), and ranges from 0 (uniform) to 1 (extremely heterogeneous). It is not a unique measure of reservoir heterogeneity since several different permeability distributions can yield the same value of Lorenz coefficient

3. Dykstra-Parsons permeability variation factor[5] is based upon the log normal permeability distribution (Figure 3-16). Statistically, it is defined as:

$$V = \frac{\bar{k} - k_\sigma}{\bar{k}} \qquad \text{(D-9)}$$

where:

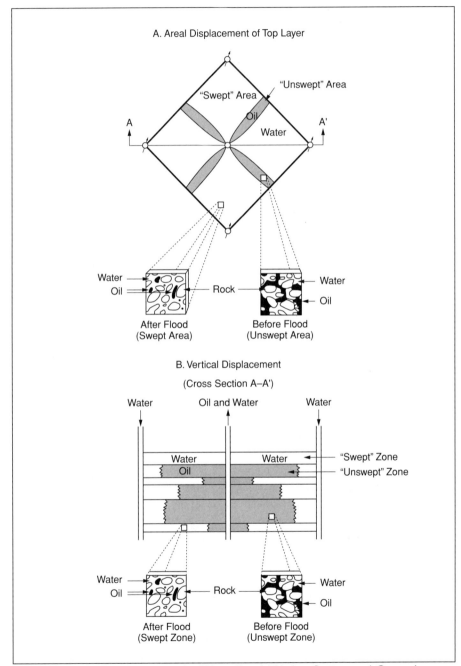

Figure D-5 Oil Displacement by Waterflood (Abdus Satter and Ganesh Thakur, *Integrated Petroleum Reservoir Management: A Team Approach*, Copyright PennWell Books, 1994)

$\bar{k}$ = mean permeability, i.e., permeability at 50% probability

k_s = permeability at 84.1% of the cumulative sample

The permeability variation ranges from 0 (uniform) to 1 (extremely heterogeneous), and is widely used to characterize reservoir heterogeneity.

Recovery Efficiency

The overall waterflood recovery efficiency is given by (see Figure D-5):

$$E_R = E_D \; x \; E_v \qquad\qquad (D-10)$$

where:

E_R = overall recovery efficiency, fraction or %
E_D = displacement efficiency within the volume "swept" by water, fraction or %
E_v = volumetric sweep efficiency, the fraction of the reservoir volume actually swept by water, fraction or %

Displacement efficiency (Equation D-8) which is influenced by rock and fluid properties, and throughput (pore volumes injected) can be determined by 1) laboratory core floods, 2) frontal advance theory, and 3) empirical correlations.

Laboratory core floods, ideally using representative formation cores and actual reservoir fluids, are the preferred method for obtaining of S_{or} and E_D). Fractional flow theory can be used to estimate Sor and ED, but it requires measured water-oil relative permeability curves. Alternately, empirical correlations such as Croes and Schwarz[6] based upon the results of laboratory waterfloods can be also used (Figure D-6).

Volumetric sweep efficiency is defined by:

$$E_V = E_A \; x \; E_I \qquad\qquad (D-11)$$

where:

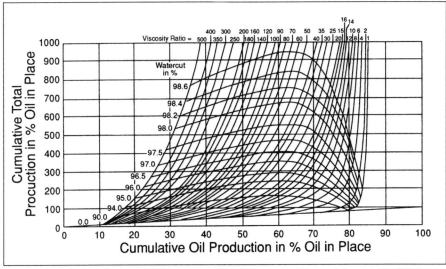

Figure D-6 Experimental Waterflood Performance (Copyright 1955, SPE, from Trans. AIME, 1955[6])

E_A = areal sweep efficiency - fraction of the pattern area or intended flood area that is swept by the displacing fluid, i.e., water

E_I = vertical or invasion sweep efficiency - fraction of the pattern thickness or intended thickness that is swept by the displacing fluid, i.e., water

The factors which determine areal or pattern sweep efficiency are the flooding pattern type, mobility ratio (defined below), throughput and reservoir heterogeneity.

Mobility ratio is defined as

$$M = \frac{\lambda w \ in \ the \ water \ contracted \ portion}{\lambda o \ in \ the \ oil \ bank}$$

$$= \frac{\dfrac{k_{rw}}{\mu_w}}{\dfrac{k_{ro}}{\mu_o}} \qquad \text{(D-12)}$$

where:

λ = mobility = k/m
k_r = relative permeability
μ = viscosity, cp
w,o = subscripts denoting water and oil, respectively

The relative permeabilities are for two different and separate regions in the reservoir. Craig[1] suggested calculating mobility ratio prior to water breakthrough, i.e., k_{rw} at the average water saturation in the swept region, and k_{ro} in the unswept zone.

Areal sweep efficiency for various patterns has been studied using both physical and mathematical models. For the five-spot pattern, the most frequently used correlations are those by Dyes, Caudle, and Erickson.[7] Figure D-7 presents areal sweep efficiency correlated with mobility ratio, M, and water cut as a fraction of the total flow coming from the swept portion of the pattern, f_D. In Figure D-8 areal sweep is related to mobility ratio and the displaceable volumes injected. The displaceable volume, V_A is defined as the cumulative injected water as a fraction of the product of the pattern pore volume and the displacement efficiency of the flood. It should be noted in the correlations that the areal sweep efficiency increases after water breakthrough for all mobil-

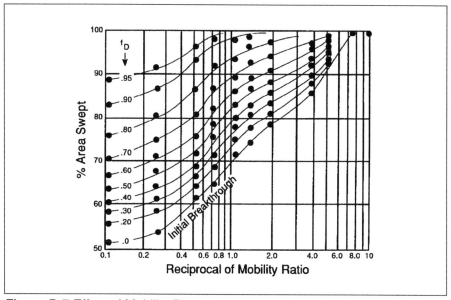

Figure D-7 Effect of Mobility Ratio on Water Cut for the Five-Spot Pattern (Copyright 1955, SPE, from Trans. AIME, 1955[7])

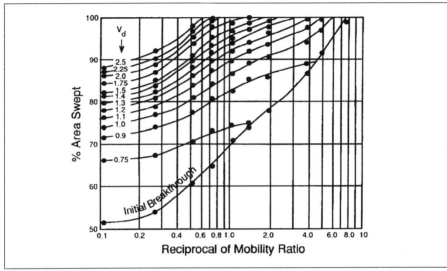

Figure D-8 Effect of Mobility Ratio on the Displaceable Volumes Injected for the Five-Spot Pattern (Copyright " 1955, SPE, from Trans. AIME, 1955[7])

ity ratios. For favorable mobility ratios, i.e., M - 1, 100% areal sweep efficiency can be obtained with prolonged injection of water.

Vertical (or invasion) sweep efficiency is influenced by reservoir heterogeneity, mobility ratio, cross-flow, gravity and capillary forces. Reservoir properties such as permeability, porosity, pore size distribution, wettability, connate water saturation and even crude oil properties can vary significantly. The variations can be areal and vertical. Permeability variation has the greatest influence on vertical sweep efficiency. Horizontal permeabilities vary with depth due to change in depositional environments and subsequent geologic events. The injected water moves preferentially through zones of higher permeability. In a preferentially water-wet rock, water is imbibed into the adjacent lower permeable zones from the higher permeable zones because of capillary forces. Also, injected water tends to flow to the bottom of the reservoir due to gravity segregation. The net effect of these factors is to influence the vertical sweep efficiency of a waterflood project.

Injection Rates

The rate of oil recovery and therefore the life of a waterflood depends upon water injection rate into a reservoir. The injection rate which can vary throughout the life of the project is influenced by many factors. The variables affecting the injection rates are: Rock and fluid properties, areas and fluid mobilities of the swept and unswept regions, and the oil geometry, i.e., pattern, spacing and wellbore radii.[7]

During the early period of injection into a reservoir depleted by solution gas drive, the water injection rate declines rapidly from an initial value. For a steady-state, incompressible flow in a homogeneous radical system, Muskat,[8] and Deppe[9] provided injection rate equations for regular patterns with unit mobility ratio and no free gas saturation. Craig[1] presented a table of these equations in his monograph.

The water injectivity is defined as the injection rate per unit pressure difference between the injection and producing wells. A drastic decline in water injectivity occurs during the early period of injection into a reservoir depleted by solution gas drive (Figure D-9). After fill-up, the injectivity variation depends upon the mobility ratio. It remains constant in the case of unit mobility ratio, increases if M > 1 (unfavorable) and decreases if M < 1(favorable).

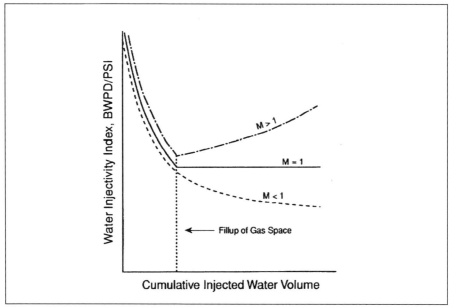

Figure D-9 Water Injectivity Variation in a Radial System (Copyright 1954, SPE, from Trans. AIME, 1954[1])

References

1. Craig, F. F., Jr. "The Reservoir Engineering Aspects of Water-flooding," SPE Monograph 3, Richardson, TX (1971).
2. Satter, A. S. and G. C. Thakur. "Integrated Petroleum Reservoir Management: A Team Approach," *PennWell Books,* Tulsa, OK (1994).
3. Buckley, S. E. and M. C. Leverett. "Mechanisms of Fluid Displacement in Sands," *Trans. AIME* (1942) 146, 107-116.
4. Welge, H. J. "A Simplified Method for Computing Oil Recovery by Gas or Water Drive," *Trans. AIME* (1942) 146, 107-116.
5. Dykstra, H. and R. L. Parsons. "The Prediction of Oil Recovery by Waterflooding," Secondary Recovery of Oil in the United States, 2nd ed., API (1950) 160-174.
6. Croes, G. A. and N. Schwarz. "Dimensionally Scaled Experiments and the Theories on the Water-Drive Process," *Trans. AIME* (1955) 204, 35-42.
7. Dyes, A. B., B. H. Caudle, and R. A. Erickson. "Oil Production After Breakthrough as Influenced by Mobility Ratio," *Trans. AIME* (1954) 201, 81-86.
8. Muskat, M. "Physical Principles of Oil Production," McGraw-Hill Book Co., Inc., New York (1950).
9. Deppe, J. C. "Injection Rates - The Effect of Mobility Ratio, Area Swept, and Pattern," *Soc. Pet. Engr. Jour.* (June, 1961) 81-91.

APPENDIX
E

Other Items Of Interest In Surveillance

S$_{OR}$/ROS Determination

Table E-1 describes tools and techniques to determine residual oil saturation (S$_{or}$) and remaining oil saturation (ROS). In using these techniques, one should keep the following in mind:

- Each measurement technique has different capabilities and limitations
- Use several measurement techniques to establish confidence in the final answer
- In open-hole environments, the NML (nuclear magnetism logs) Mn EDTA logging method yields the best accuracy
- In cased-hole environments, the PNC (pulsed neutron capture) Log-Inject-Log method yields the best accuracy

Injection and Production Logging

These logs are generally run cased hole and used to evaluate fluid flow in and around the wellbore, e.g., coning, channeling, premature breakthrough, vertical aquifer migration. The logs are quite accurate for single-phase profiling, but they are still being perfected for multiphase flow.

Figure E-1 illustrates three regions investigated by these logs: wellbore, near wellbore, and reservoir. Temperature, pressure, and trac-

Table E-1 S$_{or}$/ROS Measurement Techniques

Method	Advantages	Disadvantages
Material Balance	Straightforward calculation Produced oil usually known accurately	ROS seldom known accurately Multi-zone completions complicate allocation Areal reservoir distribution seldom known accurately
Cores		
Conventional	Direct measurement Services widely available	Core bleeds fluids when pressure released and dis solved gases expand
Pressure	Maintains reservoir pressure, thereby pre- venting oil from escaping	Expensive Poor recovery Special labs required
Sponge	Preserves oil expelled while pressure drops	Potentially more core jamming and shorter barrel than conventional core
Core Flood	Direct measurement	Must preserve wettability Core may not be representative of reservoir Core may be damaged

Table E-1 S$_{or}$/ROS Measurement Techniques (Contd.)

Method	Advantages **Open Hole Logs**	Disadvantages
Resistivity	Inexpensive Can measure either invaded or uninvaded zone	Many sources of error in interpretation
High Frequency	Inexpensive	Many sources of error in interpretation
Dielectric Logs	Straightforward interpretation Can run in variable formation water salinities	Very shallow depth of investigation; sensitive to borehole rugosity
Low Frequency	Inexpensive	Not accurate with current interpretation techniques
Dielectric Logs	Operates in very fresh formation waters	Only works in high formation resistivities
Nuclear	Measures oil directly	Mud additives may affect mud properties
Magnetism Logs	Most accurate openhole wireline technique now available	Must have complete filtrate invasion
Pulsed Neutron	Inexpensive	Depends on rock properties and brine salinity
Capture Logs	Bulk volume measurement	Sor method requires uniform filtrate invasion

Table E-1 S_{or}/ROS Measurement Techniques (Contd.)

Method	Advantages Cased-Hole Logs	Disadvantages
Pulsed Neutron Capture Logs Log-Inject-Log	Best accuracy of cased -hole techniques	Must have uniform brine displacement within depth of investigation of tool Accurate porosity necessary Expensive technique requiring special wellsite procedures
Pulsed Neutron Capture Logs Time Lapse	Eliminates uncertainty due to matrix variations	Accurate porosity necessary Only measures changes in saturation unless saturation is accurately known at time baseline log is run
Pulsed Neutron Spectral Logs	Salinity independent measurement	Measurement lacks precision Expensive Slow logging speed

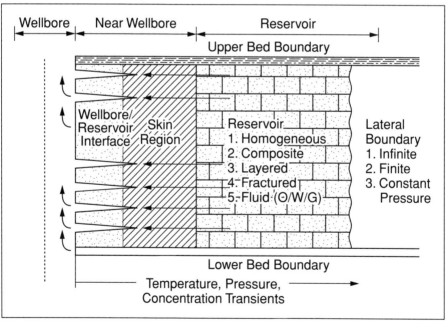

Figure E-1 Production Log Flow Regions

er signals are monitored when the reservoir is investigated. Temperature transient signals indicate near-wellbore phenomena, while pressure transient signals indicate more distant reservoir properties.

Figure E-2 illustrates logging considerations and related problems. Before designing a survey, problems are often anticipated. They may either relate to reservoir surveillance or problem well evaluation. Reservoir surveillance logs monitor well and reservoir performance. Surveys normally involve logging injection, production, and observation wells (see Table E-2). Multiple logs are often run for monitoring flood operations. Profiles determine injection and production allocation, and they relate to:

- Production and injection allocation
- Balance voidage and injection — zones/patterns
- Maintain pressure
- Detect thief zones and high permeability channels
- Locate injected fluid breakthrough
- Monitor fluid fronts

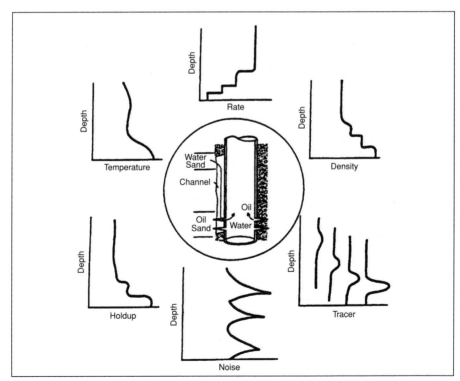

Figure E-2 Solving the Problem

- Detect crossflow and fluid migration
- Assist in reservoir simulation studies
    - water entry
    - high permeability/porosity intervals
    - history matching

Table E-2 Reservoir Surveillance Objectives

- Production and Injection Allocation
- Balance Voidage and Injection ⟵ Zones
- Maintain Pressure ⟍ Patterns
- Detect Thief Zones and High Permeability Channels
- Locate Injected Fluid Breakthrough
- Monitor Fluid Fronts
- Detect Crossflow and Fluid Migration
- Assist in Reservoir Simulation Studies
    - Water Entry
    - High Permeability/Porosity Intervals
    - History Matching

Problem well analysis is a major production logging objective. Surveys are used to resolve downhole problems such as leaks and unwanted fluid entry. Table E-2 summarizes objectives, and Figure E-3 shows a variety of behind pipe communication scenarios. In addition, evaluation surveys (see Tables E-3 and E-4) monitor and evaluate stimulation, water shut-off, profile control, and recompletion. Profile surveys are run before and after stimulation, termed as profile-stimulate-profile (PSP).

Temperature, spinner, and tracer logs are often combined in one survey to accurately profile injection wells: spinner measures high rates, tracer low rates, and temperature fluid entry and exit. Production log sensor measurements, according to the parameters they measure, are shown in Table E-5. Tables E-4 and E-5 show various production logging tools and applications.

Table E-6 provides a general guideline for preparing a production/injection log program. Quantitative and qualitative reliability of results depends on quality control, and Table E-7 describes these items in detail. Tables E-8 and E-9 show log sensors used in profiling wells. Note that the number of sensors used increases as the number of phases increase, i.e., one-phase to two and three. Figure E-4 highlights several spinner flowmeter procedures, and it can also be used as flowmeter type curves. Figure E-5 shows a similar diagram for radioactive tracer logs.

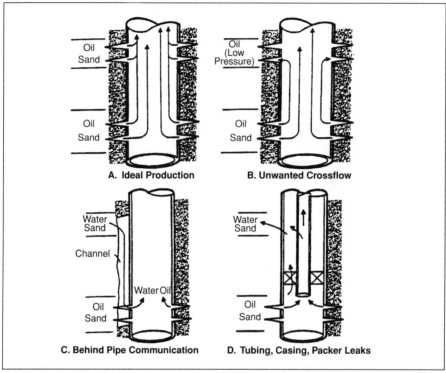

Figure E-3 Unwanted Fluid Entry Examples

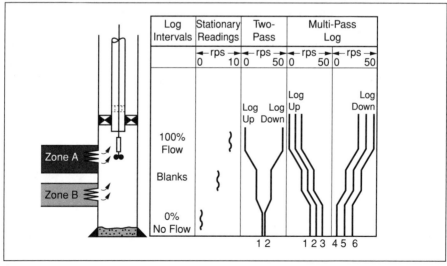

Figure E-4 Production Log Procedures

Table E-3 Problem Well Evaluation Objectives

- Mechanical Integrity Tests
  — Leak Detection

  {
  Tubing
  Casing
  Packer
  Bridge Plug
  Top Perforation
- Locate Channeling Behind Pipe
  — Water and Gas Into Production Zone
  — Prevent Groundwater Contamination
- Detect Unwanted Zonal Communication

Table E-4 Oil Field Operations Monitored Using Evaluative Surveys

- Well Stimulation
  – Acidizing
  – Fracturing
  – Chemical (Solvent)
- Mobility Control (Profile Improvement)
  – Chemical
  – Polymer
  – Foam
- Stage Treatments
  – Zonal Diversion of Fluids
  – Frac Jobs
- Water Shut-off
  – Chemical
  – Mechanical
- Recompletion

Table E-5 Production Log Sensor Measurements

Flow Rate	Fluid Density	Completion
• Flowmeters	• Gradiomanometer	• Casing Collar Locator
– Basket		
– Spinner	• Gamma Ray Density	• Casing Inspection
	• Capacitance (Holdup)	• Cement Bond Log
• R/A Tracers	• Trap	• Borehole Televiewer
• Noise	• Pressure Sensors	
• Temperature		

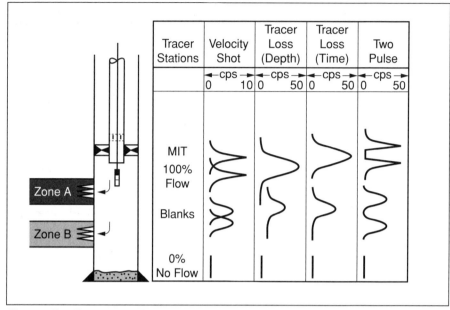

Figure E-5 Tracer Log Schematic Diagram

Table E-6 Production/Injection Log Survey Design

- Define Test Objectives
  - State the Problem(s)
  - Identify Symptoms
- List Logging Sensors Required
- Prepare Logging Procedure
- Conduct Survey
- Well Site Quality Control
- Interpret Survey
  - Quantitative
  - Qualitative

Table E-7 Quality Control Items

- Surface and subsurface calibration of logging sensors
- Surface rate measurement before and during logging
- Centralize tool string
- Reference stops in sump and blank above top perforation (100% total flow and 0% no flow)
- Depth control (casing collar locator, gamma ray)
- Multiple passes (up and down)
- Stationary readings (blanks and sump)
- Leak checks

Table E-8 Production Logging Tools and Applications

Log Sensor	Profiling	Detect Channeling	Detect Communication	Wellbore Fluid Migration	Fluid or Gas Entry	Stimulation Evaluation	Identify Fluids Flowing	Cement Location and Quality	Evaluate Tubing, Casing and Open Hole Conditions	Saturation Determination Behind Casing	Depth Control
Mechanical Flowmeter	X	X	X	X	X	X					
Radioactive Tracer	X	X	X	X	X	X		X	X		
Temperature	X	X	X	X	X	X		X	X		
Noise	X	X	X	X	X			X	X		
Fluid Density	X			X	X		X				
Capacitance	X			X	X		X				
Pressure	X	X			X	X	X	X	X		
Oxygen Activation	X	X	X				X	X			
Casing Inspection									X		
Cased Hole Nuclear										X	
Casing Collar Locator											X
Televiewer	X			X	X		X		X		X
Cement Bond			X					X			

Table E-9 Profile Surveys

Single Phase	Two Phase	Three Phase
• Temperature (E, X, C, Q)	• Temperature Pressure (E, X, C, ID)	• Temperature Pressure (E, X, C, ID)
• Spinner (Q, XF)•	• Full Bore Basket (Q_T, Mix)	• Basket/Diverter (Q_T, Mix)
• Tracer (Q, XF, MIT)	• Density (HU, ID)	• Density (HU, ID)
	• Tracer (Q_i, V_s, MIT)	• Capacitance (HU, ID)
		• Tracer (Q_L, Q_G, MIT)

Notation:
E –	Entry	ID –	Fluid Props	Mix –	Homogenize Flow
X –	Exit	XF –	Cross Flow	Vs –	Slip Velocity
C –	Channel	MIT–	Mech Integrity		
Q –	Rate	HU –	Holdup		

Once the production/injection logs are run, it is critical to document these in detail, along with why they are run, so that it is easy to utilize this information in later studies. A summary of information to capture is described in Table E-10. Also, a summary of production log survey design is shown in Table E-11.

Table E-10 Production Log Summary

- Specify Logs Run and Their Chronology
- Give Logging Sequence and Number Passes
- Specify Interval Logged and Completion
- Give Logging Speed and Direction
- Detail Calibration Procedures
  - Surface
  - Bottomhole
- Reference Crosscheck Logs
  - Relogged Intervals
  - Combination Sensors
- Summarize Results

Table E-11 Production Log Survey Design

- Define Test Objectives
  - State the Problem(s)
  - Identify Symptoms
- List Logging Sensors Required
- Prepare Logging Procedure
- Conduct Survey
- Well Site Quality Control
- Interpret Survey
  - Quantitative
  - Qualitative

Interwell Tracers for Waterflood Asset Management

Interwell tracers are used to track injection fluid from injection wells to production wells. The produced tracer timing and concentration can be used to deduce reservoir properties, as they relate to preferential flow paths (see Figure E-6). The use of tracers to obtain some relevant information is summarized in Table E-12.

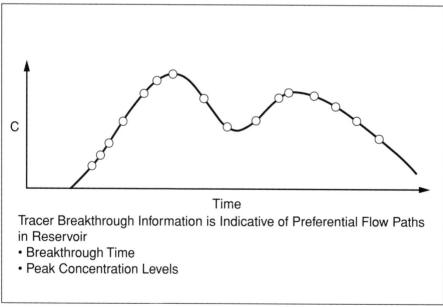

Time

Tracer Breakthrough Information is Indicative of Preferential Flow Paths in Reservoir
- Breakthrough Time
- Peak Concentration Levels

Figure E-6 Tracer Concentration History

Table E-12 Information From Interwell Tracer

- Continuity of Sands and Shales
- Characterization of Faults
- Delineation of Flow Barriers
- Directional Flow Trends
- Volumetric Sweep Efficiency at Breakthrough
- Pattern Balancing
- Identification of Problem Injectors
- Delineate Between Coning and Channeling

The continuity of sands and shales can be characterized by selective injection of tracer into a given zone (see Figure E-7). For example, if tracer is injected in only one zone and it is observed in a neighboring

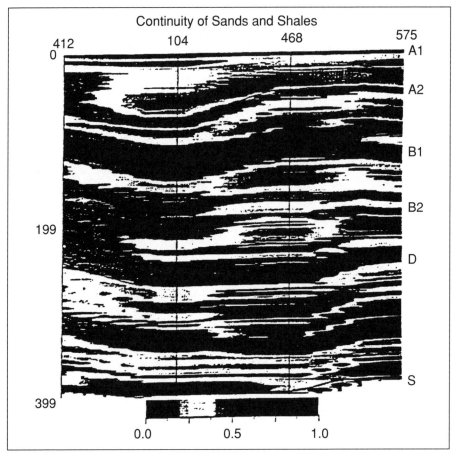

Figure E-7 Continuity of Sands and Shales

producer completed in a different zone, then the shale separating these zones is not continuous between the wells.

The absence of tracer production at an offset producer may be as a result of a fault between the injector and producer (see Figure E-8). The transmissibility in the transverse direction to a fault is generally much higher than the one along the fault gouge zone, causing a high degree of flow along the fault. Thus, early tracer breakthrough may be detected at a producer near the fault. Any type of flow restriction, e.g., low permeability region between an injector and producer, will reduce tracer movement to the producer (see Figure E-9). On the other hand, if localized fractures are present in the reservoir, early tracer breakthroughs may occur.

Preferential flow directions can be ascertained by monitoring tracer breakthroughs times at neighboring production wells in different directions from the injector (see Figure E-10). Breakthrough times are combined with pressure drops between wells to calculate the transmissibility. If preferential flow directions are present, sweep efficiency could be improved by adjusting the pattern and/or the rates. The volumetric sweep efficiency at breakthrough is a measure of the seriousness

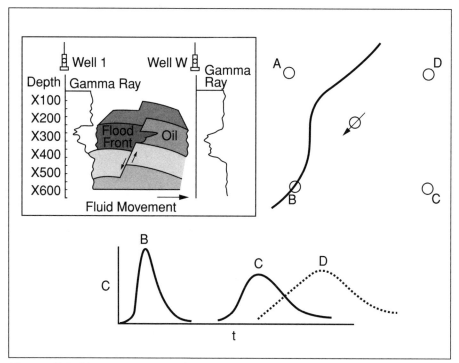

Figure E-8 Characterization of Faults

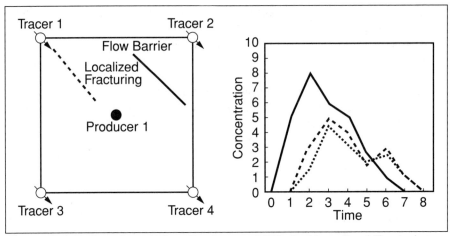

Figure E-9 Delineation of Flow Barriers and Effect of Localized Fracturing on Flow

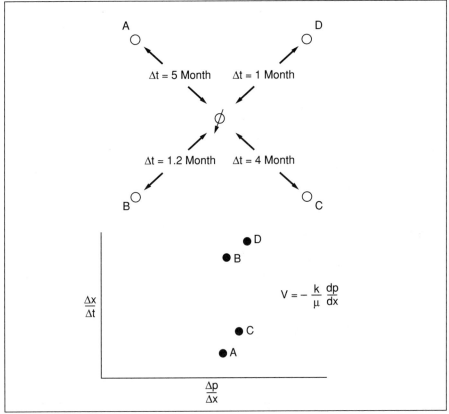

Figure E-10 Directional Flow Trends

of channeling (see Figure E-11), because the amount of fluid injected towards a producer at the time of breakthrough provides some qualitative information on the volume of the high permeability channel. Pattern balancing is important in maximizing the sweep efficiency and ultimate recovery from a waterflood. The relative amount of tracer recovered at each well provides an estimate of how much of the injected fluid flows towards each producer (see Figure E-12).

Problem injectors can be identified by tagging the injected fluid at each injector with a different tracer (see Figure E-13). Early breakthrough at a producer can be attributed to flow from a given injector. Water production at a well can be due to coning or channeling. The channeling of the injected fluid can be detected by tagging this fluid with a tracer.

Table E-13 summarizes some items to consider in a tracer test design, and Table E-14 documents criterion for tracer selection. Table E-15 describes some chemical tracers for waterflooding and their applicability.

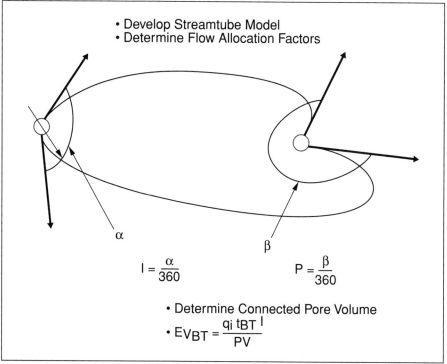

Figure E-11 Computation of EVBT (Volumetric Sweep Efficiency at Breakthrough)

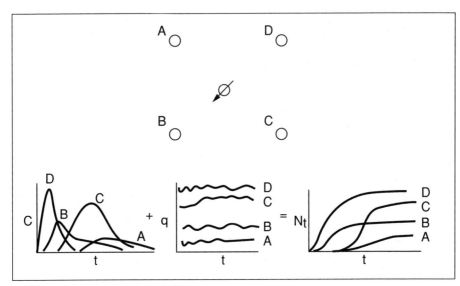

Figure E-12 Pattern Balancing

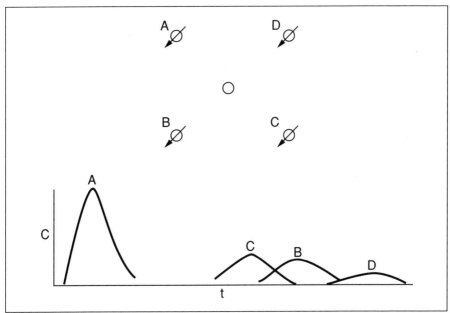

Figure E-13 Identification of Problem Injections

Table E-13 Tracer Test Design

- Define Objectives of tracer test design, e.g.
  - Lack of tracer breakthrough may be attributed to poor reservoir continuity
  - Flow boundary between injector-producer pairs
- Use of a stream tube simulation study to model the behavior of tracer injection
- Selection of a tracer
- Amount of tracer required and when to start/end injection
- Sampling frequency required and when to discontinue
- Specifications of which wells to inject in and which wells to sample
- Specification of field operating conditions during the test

Table E-14 Tracer Selection

Criterion for tracer selection requires the tracer to be:
1. Safe
2. Inert and nonreactive with the formation (rock) or fluids
3. Nonadsorbing
4. Thermally stable
5. Present in low levels in formation fluids and injected fluid
6. Detectable at low concentrations
7. Able to follow the fluid of interest
8. Inexpensive

Both chemical and radioactive tracers are used to tag injected water in waterflooding. Generally radioactive tracers are more cost effective than chemical tracers.

Table E-15 Chemical Tracers

Chemical Tracers for Waterflooding	Examples	Applicability
• Dyes	Rhodamine, Fluorescein	Detected at low concentrations (ppb), but are easily adsorbed and degraded. Recommended for minimum rock exposure and short transit time
• Inorganic Salts	Sodium, Potassium, Ammonium Bromide, Iodide, Nitrate, Thiocyanate	Nitrates go through thermal and bio-degradation. Thiocyanates also degrade at temperatures approaching 200 °F. Bromides and iodides are generally applicable, and in most produced waters they are present in low concentrations, ±5 ppm. Bromides are much less expensive (40 times) than iodides

Index

E

Economic criteria, 250-252
  Discounted cash flow return, 252
  Investment efficiency, 252
  Payout time, 251
  Present worth net profit, 251
  Profit-investment ratio, 251
Economic evaluation, 5, 15, 95, 120, 124, 127-128, 135-137, 145-146, 166, 249-260, A-304 - A-306
  Economic criteria, 250-252
Elk Basin, Madison field, 44, 261-265
  Case study, 261-265
  Elk Basin anticline, 261-262
Empirical analysis, 10, 16, 24, 163, 166-167
Encroachment, natural, 224
Environmental regulations, 202
Equipment design, 144-145
Example, data acquisition/analysis, A-347 - A-354
Example, waterflood design, 147-161
  Production rates, 155-161
  Reserves forecasts, 155-161
  Reservoir data, 148-154
  Reservoir modeling, 153-156
Exploration/development integration, 54-59
Exponential/constant decline, 172-173

F

Facies maps, 25, 34, 36-38
Failure analysis, 18, 146-147, A-315 - A-316
Feasibility study, 127-128
Field automation, 297
Field data, 80-85

Profile surveys, 82-85
Tracer surveys, 84
Water injectivity, 81-82
Well pressures, 82-83
Field discovery/development, 266-272, 279-287
  Denver Unit, 266-269
  Jay/LEC fields, 279-286
  Means San Andres Unit, 270-272
  Ninian field, 287
Field operations, 201-247
  Conversion vs. newly drilled wells, 241-245
  Injectivity increase, 245
  Subsurface/surface fluid control, 223-240
  Thought items, 240-241
  Water system, 201-223
Fines migration, 208-209
Fingering, 18
Flash liberation test, 64, 66-67
Flood pattern, 10-11, 15, 30, 45, 87-88, 93, 115-121, 131, 133, 141-143, 155-156, 168, 189, 243, 252, 267, 274-276, 280-282, 284-285, 287, A-362 - A-365, A-387 - A-388
Flood response, 16, 20
Flood-front map, 185-187
Flow capacity, 48
Fluid control, 191, 221, 223-240, 296
  Treatments, 224-228
Fluid distribution, 12, 14-19, 39, 82, 123, 135, 140
Fluid properties, 3, 10, 12, 14-19, 24, 28, 39, 63-73, 87-88, 135, 138, 166, 183, 271, 288
  Transport properties, 138
Fluid saturation, 14-19, 140
Formation damage, 80, 208-209, 222, 234-235

Performance evaluation, A-307 - A-308
Performance monitoring, 183-199, A-306 - A-307
Performance prediction, 180-181
Permeability variation, 12, 96-100
Permeability, 11-12, 14-15, 17-18, 20-25, 31, 40-42, 45, 48-50, 53, 65, 69, 75-78, 80, 83, 87-89, 92-93, 96-100, 104-108, 138-139, 149, 153, 165, 167, 189, 226, 231, 263
Permeability/thickness map, 42
Petrochemical Open Software Corporation (POSC), 55
Phase equilibrium, 138
Phases, development, 127-133
Physical characteristics, reservoir, 185
Pilot tests, 133-134, 142, 144, A-324 - A-331
Planimeter, 41-42
Plugging treatments, 225-232
Polyacrylamide gels, 231-232
Polymer systems, 230-232
Pore volume, 95, 99, 100, 103-104, 107-109, 111-112
Porosity, 14, 22, 24, 31, 39-40, 42, 65, 83, 89, 138, 152, 165
Porosity/thickness map, 42
Prats-Matthews-Jewett-Baker method, 167-169
Precipitation, 218-219
Present worth index, 252
Present worth net profit, 251
Pressure transient analysis, 82-83, 139-140
Pressures, 74-76, 82-83, 139-140
  Capillary, 74-76
  Pressure transients, 82-83, 139-140
  Well, 82-83
Primary recovery, 14-15, 88-94, 96, 185, 273-276

Probability, 89
Problem areas, wells, 189
Process, reservoir management, A-300 - A-308
Process/operations design, 141-144
  Injected fluid, 143
  Pattern type, 141-143
  Project duration, 143
  Reservoir performance, 144
Produced water, 189, 205-206, 210
  Analysis, 189
Producing mechanisms, 1-2
Production curves, 185
Production data, 28-30, 45
Production performance/reserves forecast, 93, 120, 163-182, 202, 292
  Classical methods, 167-169
  Empirical methods, 166-167
  Performance curve analyses, 170-178
  Reservoir simulation, 177-181
  Volumetric methods, 164-166
Production profiles, 82, 85
Production rates, 113-114, 155-161, 166, 170-178, 285-286
Production wells, 133, 244
Productivity, 20, 143
Profile control, 191
Profile modification, 225-232
Profile surveys, 82-85
Profiles, 82-85, 191, 225-232
  Control, 191
  Modification, 225-232
  Surveys, 82-85
Profit/profitability, 9, 251-252
  Profit-investment ratio, 251
  Profitability index, 252
Project duration, 141, 143
Project economics, 249-260
  Data, 252-253
  Economic criteria, 250-252
  Economic evaluation, 253-258